the gut stuff

First published in the United Kingdom in 2021 by
Pavilion
43 Great Ormond Street
London
WC1N 3HZ

ISBN 978-1-91166-347-8

A CIP catalogue record for this book is available from the British Library.
10 9 8 7 6 5 4 3 2 1

Reproduction by Rival
Printed and bound by Toppan Leefung Pte. Ltd.
www.pavilionbooks.com

Publisher: Helen Lewis
Commissioning editor: Sophie Allen
Infographics design: Scarlett Chetwin, Harry Lee and Myron Darlington at Revolt
Logos and branding: James Knowles Ritchie
Book design: Nathan Grace
Production manager: Phil Brown

Photography credits: p10: Josh Exell; p.41: Emma Croman; p119: Joe Pollard

DISCLAIMER: The information in this book is provided as an information resource only and is not to be used or relied on for any diagnostic, treatment or medical purpose. All health issues should be discussed with your GP and/or other qualified medical professional.

the gut stuff

an empowering guide to your gut and its microbes

by lisa and alana macfarlane
foreword by tim spector

For Dad, for giving us the guts to believe, and
Kristy Coleman, who co-wrote this with us, with
a toddler and a newborn–our hero.

PAVILION

what's inside?

(other than your gut ;))

foreword   pg 6

introduction   pg 8

 back to school   pg 12

  biology (and a dash of music)

  microbiome 101

  modern studies/politics

 mind and body   pg 38

  the gut and...

  ...immunity

  ...excercise

  ...skin

  ...brain—a tennis match (yes, really!)

  ...sleep

  ...hormones

3 scientific interlude on

pre- and probiotics   pg 64

4 bullsh*t bin (sorry mum!)   pg 74

   myth-busting toolkit

5 what can you do?   pg 82

   fermented food—is there evidence?

   simple swaps

   alcohol

   de-stress

   other sh*t you should know

6 i've gutta problem   pg 124

7 the future of science   pg 138

8 gut glossary   pg 146

foreword

As Director of the Department of Twin Research at King's College London, I have come across many thousands of twins over the years. None are like the Mac twins! Alana and Lisa have an amazing infectious enthusiasm, intelligence, and passion for research, and for disseminating that science to the public. They quickly became my go-to twin guinea pigs to road-test new research projects.

I met them around 10 years ago when they answered my call to participate on an epigenetic research project, where I was looking at why identical twins could often look identical but be quite different in many ways. They ended up being a great case study in my book, *Identically Different*, with their different personalities and gut problems. They then eagerly volunteered for more studies and went on to being the very first participants in a pilot study for our novel research project into the gut microbiome and nutrition. This study involved having all kinds of biopsies, plenty of poo samples and eating several weeks of junk food. Luckily they were performing at the Edinburgh festival, where this was readily available. They survived this ordeal followed by four weeks of healthy vegetarian high-fiber food. Their results showed us that we could alter the gut microbes with diet and allowed us to start the big study in hundreds more twins and open up the whole field.

The research evolved into the world's largest personalized nutrition study—with the help of a company called ZOE—called the PREDICT study. The twins were, once again, the first guinea pigs and had to say how they reacted differently to the identical foods, blue muffins (and prosecco). They made the point brilliantly of how unique all of us are. They both were fascinated by the science of the trillions of microbes living in our gut that are essential to our immunity and overall health, and how looking after them through a diverse and varied diet could help prevent many modern health conditions. As identical twins they are used to sharing all their genes, but suddenly they have something that is unique to them—the microbes in their gut. Through their participation in the research, they never stopped asking ever tougher and more intelligent questions about the science of the microbiome and, most importantly, they showed a unique talent not just for processing complicated scientific concepts but for translating them to a nonscientific audience in a fun and highly informative manner.

I was not surprised when they went on to found the incredibly successful The Gut Stuff, a platform to empower gut health for everyone.

This book is their next step in their efforts to make microbes, the science of gut health, and nutrition available for everyone and show that science does not have to be boring. Through their ability to attract and talk to the best experts in the field they have managed to summarize a wealth of scientific knowledge and different viewpoints to educate a young audience on what they need to do to maintain and enhance their health. I was asked on a nutrition panel recently, "What single factor is the most important in changing nutrition?" I replied, "educating everybody, even at school." As my own books, *The Diet Myth* and *Spoon-Fed*, underline, there is a real need for better nutritional health that demystifies food and nutrition. Alana and Lisa's book deserves top place in the list of books every young and not-so-young person should read, because as they themselves say, "Gut health is serious shit." Enjoy.

Tim Spector
Professor of Genetic Epidemiology, King's College London
Author of *Spoon-Fed* (2020) and *The Diet Myth* (2016).

introduction

Let's face it, talking about the gut ain't sexy. Just googling "gut" brings up a rather disturbing mosaic of beer bellies, intestinal diagrams, and, the main reason for our misconceptions, perfectly manicured hands cupping toned, soft stomachs. It's no wonder we're all confused.

So many of us have digestive issues at any given time, but we'd rather talk about ANYTHING else than our gurgling midriff and bathroom dashing. So, from this sentence on, this is where that STOPS. You are now entering an open "poo chat" forum and you wanna know why? Because it's important, really important. Hippocrates saw it many moons ago when he said:

📢 **"all disease begins in the gut"**

For some reason we've chosen to bury that knowledge. So, get your little archaeological spades and hats out, as we're about to discover what Hippocrates was on about for the good of our health. Look, nutrition is COMPLEX. Even the experts say so and, trust us, we were not experts. We had done every fad diet under the sun, including the cabbage soup diet 2005 (remember that?), and grew up in Scotland eating deep-fried pizza and fries, plus all of Edinburgh's supply of yum yums. We only knew what kale was because we used to feed it to the guinea pig on his birthday. That all flip-reversed when we volunteered to be part of the TwinsUK research at King's College London.

Being identical twins, we have a passion (teetering on obsession) for finding out what's different about us and to do this we looked inside ourselves, as there isn't much that's different on the outside. Twins are a great constant for medical research, and we became the "chief guinea pigs" for the British Gut Project. We discovered that despite having 100% of the same DNA, our guts share only 30–40% of the same microbiota, which could explain why our bodies behave so differently. And so our "gut journey" began, and now yours will, too.

Gut health is mainlining its way into public consciousness and there are lots of cool brands and products coming out that claim to help. And while some definitely do, as the category and interest grows, the cowboys start to ride into town peddling detoxes and tummy teas. Couple this with the science being pretty new (and at times conflicting), it makes for a difficult world to navigate, so KNOWLEDGE is POWER.

We've grown an expert team of scientists, dietitians, nutritionists, and doctors to keep us all on the right side of the tracks—many of whom you'll meet in these pages. Unfortunately, there isn't (and probably never will be) a magic bullet for good gut health, mainly because it's so personalized. But we're here to arm you with the FACTS, so you can make decisions that are right for you, and *disclaimer*, it's not elitist, inaccessible, or expensive.

Whether you're here because you're struggling with digestive issues, you've heard lots of chatter about 'the gut' recently and want to know what all the fuss is about, or just for the polyphenLOLs (you'll get that joke soon), welcome to the Gut Gang.

why now?

we've always had guts—so why are we all just talking about the gut now?

There's stuff we've known for a pretty long time; for example, that the gut is pretty clever and acts like our second brain, even communicating how we feel. (The gut and brain are actually working much more closely together than we thought—more on this later.) Aside from just how fascinating the gut is, with so many people encountering digestive issues, there's a huge demand to know more. Even if you don't have digestive problems, looking after your gut is still as important as looking after your heart and all your other bits and bobs. Here's a rundown of some of the main reasons we're all talking about our guts.

Looking back to 1990 there were approximately twenty-four studies on the microbiota published that year; fast forward to 2019–2020 and there were over 9,000 studies in just one year, bringing it to a total of over 40,000 in thirty years—that's a lot of research. *NOTE this includes all microbiota, ranging from the gut to the skin. Despite all this research, we only know a fraction of what our gut does and how powerful it really is.

1. First up, we're learning more about the microbiota (jump to page 20) and what our gut bugs and traveling bugs do for and against us. Many bacteria don't like to be exposed to air, so when scientists were trying to study them out of the body, they died, and so we had limited knowledge about them. But now we have the technology to study the DNA of these little guys instead of attempting to grow them in labs, which is why we are making so many advances in getting to know them better.

2. Digestive issues are on the rise; 86% of all British adults have suffered some form of gastrointestinal (GI) problem or ailment in the last year, encountering symptoms ranging from bloating and excessive wind to crippling pain and chronic disease.

3. In the Western world, we've forgotten about our gut microbes, along with our Walkmans and VCRs, and we've increased our consumption of ultra-processed foods bursting with additives and emulsifiers, which aren't great for our microbes.

4. Despite the availability and variety of unprocessed food being greater than ever, the majority of us aren't getting enough fiber (see page 90 for more), which has a huge impact on our gut microbes.

5. While antibiotics have saved many lives, used incorrectly or without the right support, they can act like a nuclear bomb on our gut microbes, affecting the balance between the beneficial and less helpful microbes and even wiping out entire species altogether.

6. We've also created a very sterile environment with our hand rubs and bleaches (rightly so at times), but in the process we've killed off some good microbes we actually need.

7. Our sterile environment coupled with more sedentary lifestyles and spending too much time in our homes may be damaging our health.

Now that we know the *why*, it's time to go back to school and learn the *what* before we get onto the *how* (this is like a game of Cluedo!).

chapter ①

back to school

biology
(and a dash of music)

so, we think it's time we got to know our guts,

and for that we need to go...

Back to School

"pencil cases and paper planes at the READY!"

Since this isn't actual school and we can do what the hell we like, we thought we'd mash up a biology and music class—where cells meet bells and drums become bums.

Our gut is a wonderful orchestra, and we don't just mean the trumpet at the end. The surface area of our digestive system is forty times larger than our skin and there are several organs involved (and we don't mean the church kind). So that's a lot of orchestra seats and stands, with Mother Nature the best conductor of all.

what is the gut?

you are what you absorb!

starts here

oesophagus

the gateway
between the outside
and your insides

supported by:
the liver
gallbladder

stomach
pancreas

small
intestine

6m*

"huh?!
yeah it's
curled up!"

"i'm only 1 cell thick"

large
intestine
1.5m*

rectum

ends here

*approximately

"what's my gut? just my stomach yeah?"

Your stomach isn't actually your gut but one player in tune with other organs, including your liver, gallbladder, and pancreas, working in harmony to support your gut. Yes, it can sound more like death metal than Bach at times, but all the players are there with their parts ready and willing it to work.

We shall call this aria, "To eat: the long journey out."

Our mouth is the perfect introduction to the show, with a plodding duet of your teeth and saliva working together to break down your food both physically (teeth) and chemically (by clever enzymes contained in your saliva). Their tunes dance around each other with the same aim: to get the food as small as possible to make it easier for the next key players to read the sheet music. But too many of us don't let the tunes play for long enough—cue not chewing—making it hard for the next players to continue playing.

side note If you thought your tongue was just for tasting, you've very much underestimated it—there are actually immune cells at the back of our tongue that are "on guard" ready to protect us. More on them later...

The esophagus is like a didgeridoo with a sphincter (valve) at the top and bottom that moves food from the back of your throat to your stomach. This is not to be confused with your windpipe (the clarinet next door). You'll know if these two have ever been confused *spluttering cough*. The sphincter at the top stops air going in and the sphincter at the bottom allows food to enter the stomach but prevents the acidic contents of your stomach going up.

And now for the stomach, the big clever bagpipe, which acts as an expensive mixer to churn our food into chyme (another musical reference nearly hit, but not quite).

Our stomach is a lot higher up than we think (see page 15), so often when people complain of stomach ache, it's not actually stomach ache; it's probably something a lot farther down the line.

Firstly, the stomach separates solids from liquids. The liquids get a VIP fast track onto the next phase, but the solids have to be mixed up in the queue a little longer.

"it's bad if our stomach is really acidic."

WRONG. Our stomach needs to have a low pH (acidic) to break down our food; the enzymes really like an acid party up in there and they need it to get rid of the unwanted 'microbes' that weren't invited.

And so, to the small intestine, where the song starts to gain momentum, as it's got a lot of pals to help the process. Nutrients are absorbed here by the villi, which we think look like sea anemones, and the food is here for 2–6 hours (it's a long tune!).

The small intestine is lined with tiny finger-like structures called villi. The inner wall of villi is only one cell thick, so it's a sensitive Sally, and it's pretty easy for substances to pass in and out of it (so not a soundproof music hall). Some foods and medications, excessive alcohol and stress may make Sally feel a bit out of sorts and she might occasionally let substances—like food particles and toxins produced by bacteria—into the bloodstream that she shouldn't.

In comes the liver, like a big important double bass (arguably the most important organ, which is why it deserves its own bullet points). The liver tends to command a lot of attention (and rightly so) because it does so many things, the key ones being:

- Produces bile acids that "skoosh" in via the gallbladder, which is a trusty sidekick that stores all the bile acids (not to be confused with hangover sickness) when the time is right—like a steady cello perhaps?

- Filters blood coming from the small intestine, in doing so it:

  - "polices" what's in the blood, such as toxins or medications, which it then has a good sort through and either turns them into something less harmful or metabolizes them.

  - converts food to fuel to give us energy (even our gut needs energy to work).

  - stores vitamins, minerals, fats, and sugar for later use.

- Produces hormones to help regulate lots of different functions around the body.

To round off the string section is the pancreas, which secretes enzymes and hormones that play a part in controlling blood sugar and sodium bicarbonate to change the pH of chyme that enters your small intestine and other things.

Hold up! Chyme? Whaaat?!

What is it? A cocktail of stomach acid, digestive enzymes, partially digested food, and water.
What does it do? Chyme allows for further digestion by enzymes and carries food and enzymes to the small intestine.

Into the plodding adagio we move, as all the unabsorbed bits of food make their way into the large intestine (for around 12–30 hours—hope everyone brought snacks to this show). Here is where most of our gut microbes live (more on this after the show), and they LOVE everything that our human enzymes can't break down, like fiber.

We forgot about the appendix. (Most people do.) It's got a reputation for being the spare part, like the cow bell. It sits just below the junction between the small and large intestine and is seemingly completely bypassed. However, just like the cow bell in the Christmas nativity (nothing else can signify donkey steps quite like a cow bell), it isn't just there for show. Some experts believe the appendix is like a big storage container for all the most helpful bacteria, ready for when we need them most, like when we have diarrhea.

AND SO FOR THE BIG CRESCENDO, aka getting it out the other end. This is our last chance to absorb any water. The music (hopefully the poo) swells as our intestinal muscles push your poo along to sit in the waiting room (your rectum)...the music stops dramatically...as our external sphincter closes, ready for the opportune time to introduce the trumpet and complete the song.

The audience are in POO-sition (pages 122–123 for how to do this optimally), the brain signals it's time for the eagle to land and the final trumpet sings the final notes: *DA DA DA DA DA AAAAAAAA!*

A round of applause, not dissimilar to the sound of a flush, and the audience are on their feet (washing their hands). *What a show!*

And if that wasn't enough, there's also a whole other neighboring orchestra waiting in the wings to help out too...enter THE MICROBIOME.

microbiome 101

to rephrase Ellie Goulding, "When I'm with ME, I'm standing with an army."

There will be a lot of music references across these pages. the by product of two DJs writing a book about science.

We are not all human. (Don't worry, you've not dipped into a sci-fi novel in which fourteen-fingered extraterrestrials escape from your intestines to conquer the world as we know it.) No, we have well over a million little critters like bacteria, viruses, fungi, and other organisms just chillin' in and around our body, mostly in our large intestine, also known as our gut microbiome. It's a bit like a tropical jungle, with loads of different species living and working in harmony. Now, if you'd told us this a couple of years ago, we would've wanted to chuck them right off our turf and out onto the street without an eviction notice. However, it turns out we need them.

Despite our microbes being scattered around the body, scientists are starting to treat them as an organ within their own right. Welcome to the stage, little bugs, your time has come to shine *in the style of Simon and Garfunkel*. Scientists are discovering that not only do our microbes outnumber our genes, but they are potentially just as influential. If this were *Game of Thrones*, they would be an army we definitely want to have on board. They are devoted to us from birth and maybe even before (thanks Mum)—even if in the Western world we've unknowingly been doing all we can to deplete them (treason!). But more on that later.

All these microbes are exceptionally clever and help to control your blood sugar; produce vitamins; manage cholesterol and hormonal balance; prevent you from getting infections; control the calories that you absorb and store; communicate with your nervous system and brain; and influence your bone strength, alongside hundreds of other functions!

Similar to the sage advice Aibileen gives Mae Mobley in one of our favorite film quotes from *The Help*, our microbes are kind, smart, and important.

kind
Microbes protect us. They know the difference between harmful party crashers like pathogens and harmless vacationers just fancying a journey through your intestines.

smart
Microbes are much more agile than our human cells. They can reshape and cultivate according to their environment (absolute ninjas). They can actually swap genes and bits of DNA among themselves, like a tiny miniature microbe stock exchange.

important
Microbes affect pretty much everything, from our ability to break down food, absorb nutrients, make vitamins, regulate appetite, and produce feel-good hormones. They can even inhibit the production of hormones that make us feel happy (blows our minds every time!).

Your gut microbiota may also play an important role in the development of Alzheimer's and Parkinson's disease, attention deficit hyperactivity disorder (ADHD), autism, diabetes, obesity, polycystic ovary syndrome (PCOS), and autoimmune conditions, and it can also influence your cardiovascular health. Most of the research to date points toward gut-microbiota-driven inflammation being key in the development of many of these conditions, but we are really just on the precipice of discovering exactly how different species of bacteria can help decrease or increase the risk of developing such illnesses.

The robustness of this community of microbes can make a huge, huge difference to our health, and the most fantastic news is that we can change it ourselves. Goodness knows we need to. In the Western world we've neglected our clever microbes by consuming processed foods bursting with additives and emulsifiers; living fast-paced, stressful lives; dedicating our brains to devices late at night, impacting our sleep; and probably not getting as much movement as we should. All of which, combined, wreak havoc on our garden of microbes, resulting in it looking more like a vegetable patch that's lost its scarecrow than a beautiful, colorful lawn.

We can't say for definite what a healthy microbiome should look like, but we do know microbial diversity is associated with health. So, we need to start taking care of our microbial friends and making new ones. It doesn't have to be drastic, break the bank, or involve weird cleanses (visit our Bullsh*t Bin for more, page 75). There are some really simple things you can do to meet new bacterial recruits and help the ones you already have to flourish. We've gut you covered in Chapter 5.

Now, we know what you're thinking: "How on earth am I going to employ an army without a job description or any idea what these microbes do?!"

Role: Epithelium protector/bad ass

Length of contract: A lifetime

Payment: Health

Description:

The mucus-producing intestinal epithelium lines our digestive tract and is just one cell thick. It is fundamental for immune function and forms the first line of defense, separating your self from your nonself. It's like a skin, which when stretched out would cover a tennis court, and our gut microbes are absolutely critical to its health. If the epithelium is not looked after properly, it can become papery and penetrable, like a colander. Substances can get through it to have a swim in our bloodstream, including endotoxins (that's toxic substances bound to bacteria to you and me), proteins, and other food components. This can cause an emergency response from our immune system—*sound the alarm*—"WE HAVE INTRUDERS!" And so, inflammation, part of your body's immune response, is triggered. Don't get us wrong, inflammation at the right time and place is a perfectly normal response, but like the guy at the bar who just won't leave, it can cause problems farther down the line if it lingers on too long (becoming chronic). Chronic inflammation is linked to metabolic disorders such as diabetes, rheumatoid arthritis, heart disease, gastrointestinal disorders, poor mental health, and many more. So, it is no laughing matter, and your intestinal epithelium plays a very important role.

Your intestinal epithelium cells get replaced every 4–5 days! So, changes can be made really quickly. High staff turnover, eh? HR nightmare!

Some bacteria can directly enhance the epithelium's function, namely *Lactobacillus plantarum* (aka a tongue twister), which can be found in fermented vegetables. So, whoever receives this role must be in possession of high quantities of sauerkraut.

Now, back to that mucus lining. What's this got to do with bacteria? *Akkermansia muciniphila* (another catchy name) is a bacteria with health-promoting benefits, which loves a bit of fasting, lives off your mucus lining, and is associated with people with a lower body mass index (BMI). If your mucus layer is damaged, say by too much alcohol (oops), it can reduce the amount of *Akkermansia* kicking about. See where we are going with this? Bacteria influences almost everything. Stress, too much alcohol, and processed food can all influence your delicate gut lining, so tune in and take control to help keep self away from nonself.

Apply within.

Our microbes perform a list of jobs long enough to overload even the most prolific job centre, but it's an army we can build and cultivate ourselves. The research is still very, very new and, in some cases, conflicting. But at some point in the future, alongside self-driving cars, jars that open with ease and a phone battery invented by the Duracell bunny, the research will be so advanced and healthcare so personalized that we'll know exactly what bacteria we need and have a lovely little probiotic and prebiotic made just for us to chug down with our morning coffee. (More on this with Dr Ruairi Robertson later, see page 142.) Until then, there are simple changes (and wonderful additions) that we'll introduce you to in this book, while the scientists are beavering away.

Right, there's something we want to clear up before we move on, as this confused us for AGES—what's the difference between the gut microbiome and gut microbiota?

in this book, we will mostly use gut microbes

microbiota vs...

microorganisms by
taxonomy (name)
in one environment

bob

jo

frank

bacteria
viruses

funghi
archaea

microorganisms in
entire habitat (gut)

and their
genetic
material

microbiome

You'll see these two terms being used interchangeably, but microbiota refers to the microbes in one environment, whereas microbiome refers to microbes in their entire habitat, including their genetic material. In the grand scheme of things, providing you aren't a scientist, it doesn't really matter—it's all gut microbes to us.

Another word we hear bandied about in the gut health world is "dysbiosis," which is basically a scientific term for your microbes being out of whack.

Here are three key reasons why this may happen:

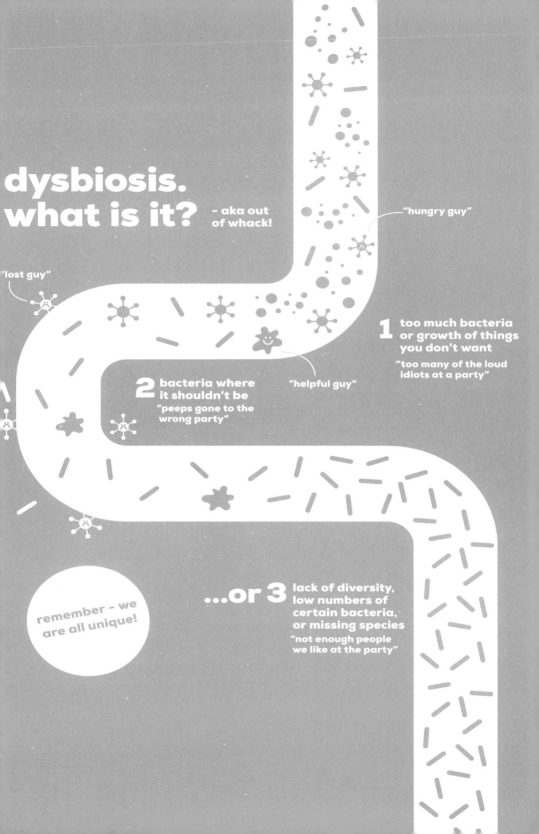

dysbiosis.
what is it?

- aka out of whack!

"hungry guy"

"lost guy"

1 too much bacteria or growth of things you don't want

"too many of the loud idiots at a party"

2 bacteria where it shouldn't be
"peeps gone to the wrong party"

"helpful guy"

remember - we are all unique!

...or 3 lack of diversity, low numbers of certain bacteria, or missing species
"not enough people we like at the party"

what influences your gut microbiome?

– quite a lot as it turns out!

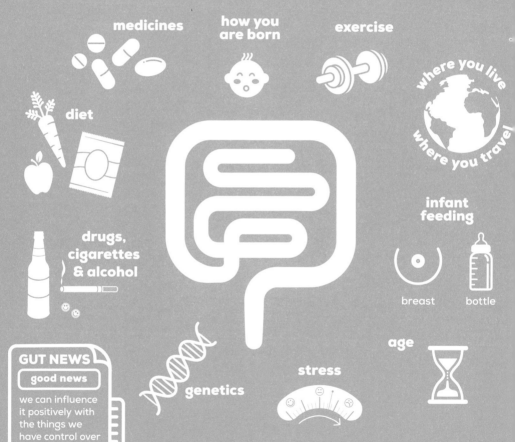

medicines

how you are born

exercise

where you live where you travel

diet

infant feeding

drugs, cigarettes & alcohol

breast

bottle

GUT NEWS

good news

we can influence it positively with the things we have control over

genetics

stress

age

OK, so we know we have these other friends. We know where they live. We know where they work. But what makes them tick and what are they affected by?

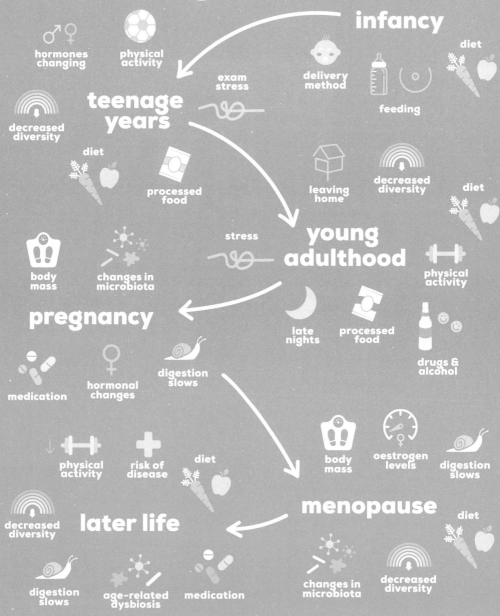

gut through lifestages

antibiotics

exposure to microbes

infancy

hormones changing

physical activity

exam stress

delivery method

diet

feeding

decreased diversity

teenage years

diet

processed food

leaving home

decreased diversity

diet

body mass

changes in microbiota

stress

young adulthood

physical activity

pregnancy

late nights

processed food

medication

hormonal changes

digestion slows

drugs & alcohol

body mass

oestrogen levels

digestion slows

physical activity

risk of disease

diet

menopause

diet

decreased diversity

later life

digestion slows

age-related dysbiosis

medication

changes in microbiota

decreased diversity

It isn't just what you eat—so many other factors play a part in the health of your gut microbes, including:

- **how you are born**

- **how you are fed
(formula vs breast)**

- **genetics**

- **age**

- **where you live/travel**

- **exercise**

- **stress**

- **drugs, medication, alcohol, and cigarettes**

- **diet**

Right, so we get that our gut microbes are quite easily influenced, which is a positive thing, as we can help them by changing factors that are within our control. But do our microbes change as we change? And age as we age?

Your gut microbes are delicate and go through lots of changes throughout your life depending on what's going on. So, no matter what age you are, you need to think about your gut.

modern studies / politics

Don't worry, this isn't a test on voting systems or the welfare state, but class and politics have been linked with nutrition and health since the beginning of time, and it's the reason we started our business. This comes down to access to information, private healthcare, accessibility of language and sources, and, of course, cash. Most of us just see health as "not being ill" and "well-being" as a middle-class world of luxury occupied by green smoothies and gong baths. In fact, it is VITAL we start to think of them together if we want to turn our focus to prevention rather than cure and help take the strain off our healthcare system.

Dr Chris George is an NHS doctor, and all his advice on lifestyle is based on scientific research, evidence, and his clinical experience as a doctor. He's also director of the British Society of Lifestyle Medicine, so there's no better person to teach this part of the lesson...

"The World Health Organization (WHO) reported that 71% of deaths worldwide are due to chronic lifestyle-related conditions termed 'non-communicable diseases,' including illnesses such as cancer, stroke, and heart disease. Chronic conditions are actually the largest cause of death and disability, but these chronic conditions are largely preventable through the modification of lifestyle factors such as diet and exercise. Diet and nutrition are important factors in maintaining good health. Despite increased awareness that poor diet is associated with negative health outcomes, food choice remains an extremely

Dr Chris George

complex subject. Within our healthcare system, socioeconomic status (SES) and education appear to be the largest determining factors when it comes to nutrition. There are numerous studies that show people's diet can be related to their SES. This means that the consumption of an unhealthy diet, classically one low in fruit and vegetables, is associated with people from poorer backgrounds. Less nutritious, energy-dense foods are typically cheaper sources of calories, whereas a higher quality diet tends to cost more. People who shop at low-priced supermarkets have been shown to have lower-quality diets and higher BMIs. Interestingly, this cannot be put down to cost alone, as people from higher-educated households shopping within the same supermarket have been shown to make healthier food selections.

What we're seeing in clinical practice is a rise in most gastrointestinal (GI) disorders, including functional GI disorders, alcohol-related disease, and rising obesity levels. Functional GI disorders, such as irritable bowel syndrome (IBS), are incredibly common and sufferers are often not seeking the help that they need from their doctor. This means that they are living with debilitating disorders, which are having a huge impact on their work and home lives. In addition, access to specialist nutritional advice is often too expensive for many patients living with chronic digestive diseases, which can exacerbate healthcare inequalities. Furthermore, with huge amounts of misinformation and health myths on social media, it's hard to find credible information from trusted sources.

To improve the largely preventable conditions, we need both a combination of individual responsibility and health-policy reforms. As this burden of chronic disease increases, there is lots that people can do to equip themselves with the knowledge and skills to make better-informed food choices. To plug the nutrition gap that has emerged within healthcare, the British Society of Lifestyle Medicine has created a platform to provide information about evidence-based interventions focused on diet and nutrition. People can access this information by looking at www.bslm. org.uk and checking out the latest events and annual conference."

Cohort studies? Case series? Anecdotal? Eh?!?

You'll see "scientific and clinical studies or evidence based" quite a few times in this book, so we thought we'd take a little tea break to teach you about the different levels and types. The "hierarchy of evidence" is basically a league table for different types of scientific research.

Systematic reviews pool all relevant studies on a topic and try to reach a general conclusion based on all the studies available, which is why they are top dog. While bias is minimized, caution and a degree of critical thinking is still needed. It may look at several small studies and pool them together, which would amplify any positive findings and magnify potential negatives. This can result in a misleading conclusion about a particular area. But when we talk about the individual studies reviewed within these reviews, RCTs are widely considered the most reliable type of single study you can do. BUT they too can have flaws...

Not only are RCTs good for monitoring what happens when you make people do something very specific, very differently, but they also RANDOMLY allocate participants to different groups for testing within the trial (intervention or no intervention)—this reduces the potential bias that may arise when selecting individuals for different groups. There are quite a few variables within this randomization and this all affects the reliability of the results. For example, do the participants know what they are receiving? The placebo effect is a real thing! That's why you will see single-blind (participants didn't know) or double-blind (neither the researchers nor the participants know) when it comes to the intervention. The participants are then following up after a set period of time, with the theory being that as the groups should be the same (on average), any differences between the groups will be due to the intervention (in theory). When looking at an RCT, always check out who has funded it (e.g. a pharmaceutical company), study size, and blinding.

Why are RCTs so important? Well, they're often called interventions, as they are unusual, as scientists actually DO something to study participants and then look at what happens once they have introduced a change (e.g. a supplement or mindfulness). Other types of research design may involve a lab experiment in controlled surroundings (so not real life) or use surveys to observe people "in the wild," and then try and draw links and conclusions between different treatments or risk factors and their outcomes. Hopefully this will make it all a bit easier to decipher and help you apply a critical eye to reviewing the evidence!

heirarchy
of evidence

systematic reviews
and meta-analyses

randomised
controlled
double blind
studies

cohort studies

case control studies

case series

case reports

animal research

in vitro ('test tube') research

ideas, editorial, opinions

chapter ②

mind and body

the gut and immunity

so, we know how the gut works, we know what affects it and how it changes...

...but what impact does our gut have on the rest of us? We've included just a fraction of the complex connections our gut has with other parts of our body and how we move, sleep, and eat. So, we are handing the mic over to specialists in their field to help you understand more.

First up on the stage, top immunologist Dr Jenna Macciochi to help explain the connection between your gut and your immune system. Jenna has a PhD, is a lecturer, and was awarded a prestigious Presidential Fellowship to combine her personal interest in nutrition with the study of the immune system and, to complete her legendary status, she also has twins (and is Scottish!).

Before this journey, we thought the immune system was basically just about common colds and packets of Echinacea. How wrong we were...

Dr Jenna Macciochi

"Most of us are aware that our immune system is our guardian against infection. But did you know our immune system has a complicated relationship with germs? Interacting with germs of all kinds—good and bad—is key to coaching and educating our immune system in how to behave properly and avoid unruly activities like allergies or autoimmune disease.

Bacteria educate our immune system from the moment we are born.
We know how important bacteria are for maintaining a normal immune system from experiments with germ-free laboratory mice born without any bacteria at all. The roots of this relationship lie in how you are born and nourished during the first five or so years of life. This is a key period of immune education that sets up your immune system for your long-term health.

Gut bacteria maintain a balanced immune system.
The immune system is complex and highly responsive to the world around us, so it's not surprising that there are many things that affect its function. And there is a good reason for this. Immunity is not hard-wired in our genes but educated by our environment and experiences. Much of this education takes place in our gut. It's true, and by now a familiar fact, that almost 70% of the entire immune system resides in the gut.

Throughout life, we are constantly exposed to new things via our gut, including foods, substances in our environment, and, in some cases, germs. But thankfully, most people have a healthy immune system that handles all of these invading objects with ease. If they didn't, it would elicit a dangerous inflammatory response every time they tried a new food or visited a new environment with different types of germs, dust, and dirt. The essential day-to-day task of your gut immune system is to maintain a balance between reacting to things that might hurt you and tolerating the things that won't, like good germs that we actually need. (Remember what we were talking about earlier—your gut keeps self and nonself apart, see page 23.) This is called 'oral tolerance.' Oral tolerance is when our immune system becomes trained to be unresponsive to something via the gut. It is extremely important to tolerate many of the harmless things we come across in our day-to-day lives, like the many components of food entering our gut. Much of this training happens in childhood, and our gut microbes help the process. Without it we can develop serious food allergies.

A diverse gut flora with many types of bacteria, fungi, and other beneficial germs is a crucial part of this process. These germs teach the cells of the immune system that not everything is bad. And what happens in the gut doesn't stay in the gut but affects the functioning of the immune system throughout the body, influencing many aspects of health, including how immune cells seek out potential cancer cells, how well we recover from infections like flu, and how well we control unruly inflammation from allergies and autoimmunity."

"A diverse gut flora is the healthiest.

And we are all unique. Most bacteria are beneficial, but some are responsible for making us unwell. An unhealthy microbiota might look different in different people, but they all have one thing in common: a lack of diversity. A diverse microbiota is more likely to bounce back from unhealthy fluctuations in diet and withstand outside intruders, and this means a much more tolerant and well-regulated immune system."

immunity and the gut

- and we ain't just talking about the sniffles!

gut microbes educate and support your immune system - between age 0–5 is crucial.

"hello and welcome to microbe middle school"

houses 70% of your immune system

dysbiosis disrupts your immune system

immune cells in your gut travel all over your body

"all aboard the bloodstream express"

inflammation is a normal immune response, but when prolonged it can impact mental health and increase your risk of developing autoimmune and inflammatory conditions as well as metabolic disorders.

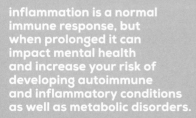

impacted by stress, pathogens, modern living, allergies, medication, and sleep

how do you keep your gut microbes (and your immune system) healthy?

Decreased exposure to infections in early life through improved hygiene was originally thought to be the main way our immunity was "trained." But rather than being too clean, the most important change in our environment that leaves us open to allergies is the loss of contact with our "old friends"–the many harmless microbes in us, on us and around us from birth. So how does our immune system see these "old friends"?

The birthing process sets off the most radical transformation–a tsunami of good gut microbes colonizes the (relatively) sterile baby as it enters the world. So, receiving foundational microbes with the potential to educate immunity starts with your mother. These initial microbes were inherited from her mother, and so on. The next thing is breast milk, which comes conveniently packaged as a food source not just for the baby but for their gut bugs too. In contrast, caesarean sections and formula feeding have been shown to have a negative impact on our gut bacterial species. While standard infant formula doesn't contain these natural milk prebiotics, some formulas now have different added beneficial prebiotics. Apart from our mothers and our diets, we also obtain our microbes from our environment. So regardless of how you are born, there is plenty you can do to change your gut microbes for the better.

You cannot control how you came into the world, but there are some things you can do to improve your gut health:

- Get more plant diversity in your diet. The fiber in plants passes through the digestive system until it reaches the colon, where it provides the fodder for your gut microbes (more on this on page 88). When they chow down fiber, they release a load of powerful immune-nourishing molecules.

- Get out, get dirty and rewild your microbes. You might have heard that we are "too clean," which is bad for our health. As well as getting microbes via our diets, we also get them from our environment. Dirt is good. Disconnection from nature is not. Modern urban life is low on microbial diversity and discourages contact with beneficial environmental microbiomes. For those of us not living in the countryside, our immune systems may be missing out on those environmental microbes. Ensure a regular counterbalance, like spending time in a garden, regularly stepping out into the countryside or a park, and doing some gardening (even if it's just your windowsill or going to a community allotment). The more contact we have with dirt and natural environments, the more we let their microbiomes infiltrate and nurture our own.

- Be mindful of antibiotics. Unfortunately, antibiotics sweep through the gut and kill off both unfriendly and beneficial gut microbes. It is best to try to avoid them where possible, but sometimes you need to take a course of antibiotics. After you've finished the treatment, the beneficial gut microbes and the unfriendly ones slowly rebuild and, if all goes well, they come back into balance. But, it takes time, and they don't always colonize in harmony. Probiotics may help your gut recover its balance faster. Consuming probiotics reduces the incidence of diarrhea associated with taking antibiotics, which affects around 30% of people. However, there aren't enough studies of any one particular probiotic to say conclusively this one works and that one doesn't. Different strains do different things. Follow the tips for general good gut health in chapter 6 (see pages 124–137).

Take your time—probably one of the biggest hurdles in our busy modern lives. Planning regular meals can be helpful for reducing stress and making sure that you don't overeat at your next meal, both of which are bad news for your beneficial microbes. Excessive stress can negatively impact your immune system too.

the gut and exercise

We know that what you eat can have a big impact on your gut. But did you know that exercise has the ability to change your gut microbes independently of anything else? Sounds like an easy win to us. Over to our head of nutrition and all-round brainbox Kristy Coleman for this one...

"Exercise can help change both the composition of your gut microbes and how they function, and that's even without having to eat some funky fermented foods. But it really depends on the type of exercise you are doing and how hard you are doing it, so it's not that simple (as you're probably starting to gather).

When the microbes in your gut ferment fiber, they produce short chain fatty acids, one of which is called butyrate. Butyrate is fuel for your gut cells and keeps your gut lining healthy, helps regulate your immune system, and supports other functions too (clever clogs!). Some studies even suggest it may play a role in sleep, supporting immune function, and helping with diarrhea control. So, in short, we want our microbes to be producing it and have the right ones there in the first place to do so. Research has shown that just five weeks of exercise increased the number of microbes producing butyrate. More research on rugby players showed that they had forty different types of bacteria compared to those who were sedentary, aka coach potatoes (scientific

Kristy Coleman

term). Studies in women have shown that those who did at least three hours of exercise per week had increased levels of certain bacteria known to produce butyrate and had high amounts of a bacteria called *Akkermansia muciniphila* (which lives in your mucus lining, see page 24), which is associated with a lean BMI.

So, does this mean you can eat what you want providing you exercise? If only it was that simple. Everybody is different, so we can't be sure what types of exercise will do what, and it's difficult to control all aspects of a study to know what is having the impact on your microbes. Sedentary people are also more likely to eat less healthily than those who are active, so that will have an impact too. BUT the overall consensus is that exercise is important for a healthy gut, whether it's hitting the dance floor or a gym class—your gut likes you to move. Just don't go too hard on high-intensity sessions, as this can cause stress, which we know has a knock-on effect on your gut and immune system."

the gut and skin

Another expert who's constantly on our speed dial is consultant dermatologist, author of *The Skincare Bible*, and all-round queen Anjali Mahto. Our conversations with her go something like this:

"Anjali, which moisturizer has SPF in it?"
"Anjali, do chemical peels really work?"
"Anjali, *insert pic* what is this on my face?"
"Anjali, we've been asked about acne again.
Please can you do another blog post?"

You can see where this is going, so we knew we had to get her take on the gut and skin. It's a super exciting area of research, so we need a measured head to walk us through it...

"It has long been recognized in medical literature that many skin disorders are more common in those with gut issues and vice versa. Data shows that rosacea is associated with small intestinal bacterial overgrowth (SIBO), and treatment of this can improve the primary skin condition. Conversely, those who suffer with inflammatory bowel disease (IBD) are at higher risk of developing inflammatory skin conditions such as psoriasis.

Anjali Mahto

So how do the gut and skin communicate with each other? While this area of research is growing rapidly, there are a number of proposed mechanisms by which the gut and skin are understood to interact:

Microbial imbalance (see dysbiosis on page 29) in the gut may cause the production of a number of molecules or metabolites (some produced by your microbes) that can access the circulation and accumulate in distant sites, such as the skin, having a negative impact.

Disturbed gut barrier (see page 23) may result in gut bacteria themselves entering the circulation and traveling to the skin. Supportive data suggests bacterial DNA of intestinal origin has been found in the blood circulation of those with psoriasis, meaning it probably got there by escaping through the wall of the gut.

Dysbiosis or gut imbalance may interfere with the immune response in skin disease by increasing the activity of certain immune cells that drive inflammation (more on inflammation later; see page 55).

Changes in the gut microbiome may directly affect or induce changes in resident microbes in the skin (or skin microbiome).

Although there are multiple pathways allowing for gut–skin interaction, there is much we still do not understand. Gut microbiota alterations have been implicated in common skin conditions such as acne, eczema, and psoriasis. However, as these ailments have multifactorial triggers for development, including interaction between genetics and the environment, it is difficult to say how gut microbiota interactions directly fit into the complex picture of disease development.

What we do know, though, is that eating well for your skin is no different to eating well for your general health. While skin disease should not be treated with diet alone where validated therapeutic options exist, there is no doubt that diet has a part to play as a piece of the wider puzzle. Diet can influence the course of skin disease, have a preventative role in disease development, and affect future health outcomes. Both short- and long-term dietary habits can alter gut microbial composition and have the potential to affect skin in both health and disease states. This may in part be achieved by sustained eating patterns incorporating:

- essential fatty acids, found most abundantly in oily fish, and in walnuts and flaxseeds for plant-based sources.

- carotenoids, found in foods like apricots, watermelon, asparagus, carrots, squash, sweet potatoes, and tomatoes (and plenty more).

- vitamin C—think berries, citrus fruits, sweet and white potatoes, and broccoli.

- vitamin E, found in foods like sunflower seeds, almonds, hazelnuts, pine nuts (need an excuse for pesto), peanuts, salmon, and avocado.

- minerals, including copper, zinc, and selenium, found in nuts, seeds, dark leafy greens, shellfish, whole grains, and oily fish.

- polyphenols, essentially lots of colorful plant-based foods.

- fermented foods (jump to page 94)."

"Early data shows there may be a role for probiotic supplementation, but it remains to be seen how effective this approach will be. As we discover and understand more about the pathways by which the gut and skin communicate with each other, we may in future be able to target certain skin conditions by manipulating the gut microbiome."

Watch this space!

gut–brain axis tennis match

To introduce this section, we have none other than the champions of the gut–brain axis themselves: John Cryan and Ted Dinan, of APC Microbiome Ireland, University College Cork, and our hero of all things "brain," Kimberley Wilson, in a game of intellectual tag-team tennis.

player(s) 1

Ted is a Cork psychiatrist. In 2019 he was ranked by Expertscape as the number one global expert on the microbiota and also listed in the top one hundred Global Makers and Mavericks. He has published over 500 papers and is co-author, with John, of the bestseller *The Psychobiotic Revolution*. John is a senior editor of *Neuropharmacology and Nutritional Neuroscience* and editor of the *British Journal of Pharmacology*, and the list is endless for the amount of peer-reviewed papers he's written and how much he's been cited. He was also included in the 2014 list of the World's Most Influential Scientific Minds—very cool!

player 2

Kimberley Wilson is a chartered psychologist and one of the best founts of knowledge we know—her book *How to Build a Healthy Brain* is a MUST-READ. She also came third on *The Great British Bake Off* (!).

Kimberley Wilson for the opening serve...

"We need to get away from the idea that the brain and the body are separate entities that simply communicate with each other. Not only does this not make any sense (how is the brain kept alive if not through the oxygen, nutrients, and energy supplied by the body?), but it has slowed down the progress of medical and psychiatric research and treatment for decades."

a strong return up next from John and Ted...

"To have 'a gut feeling' is a term familiar to most people and translates equally well in almost all languages. Given this fact, it is perhaps surprising that the view that our gut may influence our brain has only gained traction with scientists in recent times. Study of the gut–brain axis has become one of the most hotly researched areas in biology over the past twenty years. We now know that microbes within our gut have a profound effect on how our brain functions. The majority of microbes in our gut live in the large intestine. The average adult has about 1kg (2¼lb) of microbes in the intestine, which is approximately the same weight as our brains. These microbes are fed by us and, in turn, they produce molecules that our brains require. We cannot survive without our microbes and they cannot survive without us. We have evolved with microbes and would not function properly without them. It now seems strange that for so many decades we viewed the gut microbiota as consisting of commensal microbes that did us no harm but were of little benefit. How wrong we were!"

Kimberley to deploy vagus nerve shot for the next point...

"The beautiful, wandering vagus nerve is a real multitasker. Going from your brain to your gut and connecting through all of your major organs along the way, the vagus nerve is the major structural component of the gut–brain axis. But it doesn't stop there—the vagus nerve is also the main structural feature of the parasympathetic nervous system (PSNS). The PSNS is the flip-side of your 'fight or flight' response, aka your sympathetic nervous system (SNS). While the SNS is responsible for preparing your body for action—increasing blood pressure and breathing rate, shutting down digestion to make more oxygen carried in your blood available to your arms and legs etc.—the PSNS returns the body to a state of calm and rest. This is the reason that IBS symptoms are so often triggered by stress and why you should avoid trying to eat when you are agitated or in a rush; you won't be able to digest your food properly."

the microbial ball serve from John and Ted...

"How do gut microbes and the brain communicate? There are several parallel routes of communication. The long, meandering vagus nerve sends signals in both directions from gut to brain and brain to gut. Some bacteria cannot signal to the brain if this nerve is damaged. Another important communication pathway is the production by bacteria of short chain fatty acids (SCFA), such as butyrate (see pages 46–47). These are important molecules for generating energy but can also impact how the genes within our cells work. Microbes also produce tryptophan, which is the building block for serotonin (5-HT) in our brains. Serotonin plays a key role in regulating sleep and mood, and most medications used for treating depression act upon this chemical. Our research has shown that the gut microbes of those suffering from depression differ from the microbes of healthy individuals. To maintain a healthy state of mind we need to look after our gut microbes."

Kimberley's final volley...

"There is another important way that the gut communicates with the brain and that is through the immune system. The majority of the body's immune cells are found around the gut. This makes sense because food is one of the most common carriers of harmful bugs (like spoiled meat or a dodgy prawn sandwich left out for too long) and your body wants to be ready. Inflammation is the immune system's response to illness or injury. When a pathogen, like a virus or harmful bacteria, is detected, immune cells kick into action, releasing signaling molecules and powerful chemicals to kill off the threat. Once the threat is under control, the inflammation will normally die down. However, lifestyle factors such as diet can also influence immunity.

If you are not getting enough fiber in your diet, your gut microbes effectively begin to starve. Faced with this situation, they turn to a backup fuel source called mucin. Unfortunately, mucin forms the protective mucus layer that coats the inside of your gut (see page 23). If your gut microbes eat through it, the tight cell junctions in the gut wall can open and become permeable. When this happens, bacteria and food molecules from the gut can enter the bloodstream. Seeing these intruders in the bloodstream, your immune system launches an attack and if this goes on for a long time you are likely to be in a state of chronic inflammation. As this blood flows through the brain it can trigger neuroinflammation, which is associated with an increased risk of a range of mental health conditions, including depression, bipolar disorder and Alzheimer's disease. So, feeding your gut with plenty of diverse sources of fiber is actually a crucial part of a brain-healthy lifestyle."

And there we have it, a physical and chemical (and an immunity bonus) match of the best minds. Now for the post-match breakdown and takeaways.

here are some player profiles of your chemical teammates, so you know who you're working with

serotonin
your happy hormone

A staggering 95% of your serotonin is produced in your gut by your gut microbes. If you have low serotonin levels, you may experience food cravings, low mood, and even depression.

dopamine
your motivation/reward hormone

If your gut microbiota is disrupted, it can inhibit the cells that make dopamine. Low levels of dopamine are associated with low motivation, difficulty concentrating, and mood swings.

gaba
(gamma-aminobutyric acid)
your calming hormone

Your gut microbiota has the ability to increase GABA receptors in your brain, which means your brain can essentially utilize more of this chilled-out hormone. Low levels of GABA are linked to anxiety and restlessness.

here is a really common area of mental health— anxiety—at a glance...

anxiety and the gut

the brain

Amygdala may be influenced by your gut microbiome – more activity here may increase risk of anxiety.

physically connected by the vagus nerve.

neurotransmitters

bidirectional connections

chemically connected by neurotransmitters. neurotransmitters play a key role in regulating the gut.

serotonin
Happiness, linked to body clock and anxiety (if levels low). Made mostly in the gut, utilised by the brain.

the gut

gaba
Calming, moderates anxiety, made in brain and gut. Gaba receptors in brain and the gut. Low levels increase risk of anxiety.

Disrupted gut microbiome increases risk of anxiety.

Gut bacteria produce butyrate – optimal butyrate levels associated with lower risk of anxiety.

the gut stuff
×

Anxiety UK
Here for you since 1970

dopamine
Motivator, pleasure, memory, mood and sleep. produced in the gut. Low levels increase risk of anxiety.

the gut and sleep

anecdotally (no science, just us!), we noticed

that when we were touring and dj-ing with

1 a.m. set times, it wasn't great on our guts...

...so the gut and sleep is an area of research we wanted to get our noses right into.

Like all important relationships, the link between our gut microbes and sleep is a two-way street. Sleep impacts the effectiveness and function of our microbes, and our microbes impact our sleep—it can be a difficult cycle to get a handle on.

The neurotransmitters produced (remember them from the tennis match, pages 52–57) and released by your gut microbes, such as dopamine, serotonin (the precursor to melatonin—your sleep-regulating hormone), and GABA, all impact your sleep. Sleep has the ability to affect all areas of your health, including your gut, immune system, and mental health. All of this is part of the gut–brain axis—it's all linked—almost like we were made this way for a reason.

We've also recently discovered that our microbes work on our circadian rhythm (why we feel awake in the day and sleep at night). Shift work, different time zones, poor-quality sleep, and/or late or irregular bedtimes can disrupt your circadian rhythm and affect the health of your microbes and their composition. Indeed, your microbes have their own circadian rhythm and have different functions at different times of the day. Think of your microbes as hotel workers: you have your regular kitchen staff making breakfast, lunch, and dinner and then, come late evening, the room service chefs take over, often with a different menu and different roles to fulfil. So, if you suddenly start demanding lunch at midnight, you can see how your microbes might be a bit confused. This kind of explains why when you're on vacation in an actual hotel in a different time zone, you have a funny feeling for the first couple of days, which will also be affecting your gut microbes' hotel, too.

Research shows that if you have better-quality sleep, your cognitive function is increased (bye-bye brain fog!) and you are more likely to have a greater number of beneficial microbes in your gut. Caveat: we still need more information about the link between cognitive function, the microbiota, and sleep to conclusively prove this relationship.

As The Rock once said in a popular Disney movie, let's turn our attention from lessons to "takeaways."

sleep hygiene

We aren't talking a clean bed (although that can help too!), but a little routine you do every night to help set you up for consistent and optimal sleep to support you and your microbes can make all the difference. We had always pooh-poohed (you don't get away from the poo chat for long!) this before as we had such extreme night patterns, but then we started it and ZZZZZZzzzzzzzzzzzzzzz...

- Set a sleep and wake time and try to stick to them—this helps your circadian rhythm.

- Limit your use of screens in the hour or so before bed. Dim your lights and avoid those horror films. There's a setting on most phones that will automatically reduce the light from a certain time in case you forget!

- Take time to wind down from your day. Read or listen to a book, meditate, enjoy a warm bath, do some gentle stretching—whatever works for you.

- Keep your phone out of your bedroom and get an alarm clock (hello old-school!) if you rely on your phone's alarm.

- Avoid stimulants before bedtime, such as caffeine and nicotine. A note on caffeine: genetics, age, and weight will determine how quickly your body metabolizes caffeine. It has a half-life of 5–7 hours, which means it isn't just about avoiding caffeine in the evening, but also later in the afternoon too.

- Make your bedroom a relaxing place for sleep. Keep it cool and dark and minimize distractions.

- Avoid overindulging before bedtime as this can disrupt your sleep and microbes (think about the hotel workers; see page 59).

You'll find what works for you; we always find that writing our to-do lists for the next day can empty our heads a bit and leave space for the nice dreams to come in.

the gut and hormones

not only do the bugs in your gut have the power to affect your mental health...

...but they also play an important role in hormonal regulation and how much estrogen is circulating in your body at any one time.

If our hormones become disrupted by our bodies either producing or circulating too much or too little (men included!), this can disrupt our metabolism, mood, how susceptible we are to certain diseases, and our cardiovascular and bone health (which is why post-menopausal women are more susceptible to osteoporosis). Your clever microbes produce something called beta-glucuronidase, which frees up estrogen for the body to use. If you don't have enough microbes producing this or if you're producing too much, your estrogen levels will be affected, which influences body fat, metabolism, bowel movements, bone turnover, and some skin conditions. So, keeping your gut microbes in check is a great start to looking after your hormones, and we sometimes think just knowing there's a link helps you to get your head around it all!

periods and your gut

this is one of the first things we asked the scientists about—we just gotta know!

You may have noticed a change in your bowel habits before, during, or after your period (as if we didn't have enough going on in that region!). The connection between your menstrual cycle and gut is an actual thing!

The shift in hormones during your cycle can increase gastrointestinal symptoms, such as diarrhea, constipation, and bloating. Increases in hormones (such as progesterone) at certain stages of the menstrual cycle may also affect the permeability of your intestinal lining, making you more susceptible to inflammation, which may increase the risk of mood disorders such as anxiety. Symptoms can differ depending on your individual cycle, so do keep notes in a diary to help get to know your cycle and gut movements better.

TAKEAWAY
More research is needed to understand the interplay between the two, but from what we do know, period constipation/diarrhoea is definitely a thing.

chapter ③

scientific interlude on pre- and probiotics

The WHO defines a probiotic as a live microorganism which, when eaten or drunk in adequate amounts, confers a health benefit on the host (you!). Probiotics can be in food or supplement form but not all probiotics are created equal—different strains have different effects, and some might have no effect at all; it all depends on the individual. Science is still learning exactly how different strains work, so watch this space.

Foods containing probiotics include live yogurt, kimchi, sauerkraut, kefir, miso, and kombucha. Try to get a mix of different types across the course of your week; we like to experiment and make our own sauerkraut, as it's cheap and super easy. (More about how to do this later; see pages 94–98.)

Here's a little picture to put in your pocket to read the gobbledygook of supplement packaging. We've made it super simple for you.

1. Ingredients: all ingredients (not just cultures) and allergens in order of weight
2. Total Active Cell Count: colony forming units (CFU)—number of live cultures (usually at time of manufacture)
   CFU for total count but for each strain better
3. Bacterial Strains: need all 3 bits:
   1. Lactobaccilus = Genus
   2. Rhamanos = species
   3. XYZ = designation of strains. Different designations of a strain will have a different role
4. Use: how to use
5. Storage: how to store—some may need to be kept cold; others somewhere cool and dry
6. Claims: approved claims for the amount of cultures in probiotic—must be scientifically evaluated and approved by EFSA

probi-what-ics?

① ingredients:
Capsule, Live
Cultures, Allergens

**② total active
cell count:**
$10^9$5.0 Colony
Forming Units (CFU)
Each capsule:
1.25^1CFU

③ bacterial strains:
〰〰〰 ADF
〰〰〰 Z3
〰〰 〰 XY4

④ use:
Dose

best before:
00/00/00

⑤ storage:
How, out of reach
of children.
Not substitute
for varied diet.

company:
Probiotic King
& Queen

⑥ claims:
Supports xyz.

When it comes to supplements, it's very difficult to decipher fact versus expensive marketing jargon to work out which products are efficacious. We'd advise focusing on the different strains of bacteria in a product and researching a bit about what that particular strain is good for.

prebiotics:
25 years on

Now we've got probiotics covered, what about prebiotics? Prebiotics are a specific type of fiber and the food for good bacteria. Great food sources of prebiotics include onions, garlic, leeks, endive, bananas (the unripe green ones that nobody wants), asparagus, artichokes, olives, plums, and apples, plus whole grains like bran and nuts like almonds.

As a nice break in the science, we wanted to share this little anecdote with you. Professor Glenn Gibson is a microbiology professor at the University of Reading. He has worked in university or research institutes since he left home (Horden, Co. Durham) in 1979 and has published almost 500 research papers, supervised 75 PhD students, and carried out 140 research contracts. He started the research on prebiotics improving gut health and has studied people's faeces for over 30 years. He has no sense of smell ;)

Glenn is the founding father of prebiotics; we met him recently and loved hearing how these breakthroughs came about from the man behind the science. We couldn't write a book about gut health and not include his story, so grab a beer and pull up a seat at the bar with him, where this story is set...

Professor Glenn Gibson

"I can remember three things about my academic life in the 1990s...

1. Working with John Cummings in his gut group at the MRC-Dunn Human Nutrition Unit at the University of Cambridge (the other microbiologist being my sadly missed friend George Macfarlane).

2. My first PhD student being Xin Wang.

3. Not suffering the major hair loss that my subsequent working career has caused.

As part of Xin's research, I was introduced to Marcel Roberfroid. Like John, he is a real gentleman as well as an excellent scientist. Marcel was a consultant to the company who sponsored Xin's project, known as Raffinerie Tirlemontoise, then Orafti (much easier to say), now Beneo-Institute. The research was on using inulin to boost beneficial bacteria in the gut, using laboratory fermenters and two human studies. This was before the molecular revolution hit gut microbiology and I remember poor Xin having to biochemically characterize every colony that had grown on thousands of supposedly selective petri dish agars.

Marcel and I would have frequent meetings in the Scandic Crown Hotel, near Victoria, when he was in London—in the bar of course! At one of these, we chatted about how the inulin was a bit like a probiotic, in that it was changing the gut bacteriology for the better, but it did not have the survival issues that using live microbes in the diet could have. We decided to write a review on the research, and other similar studies, showing how carbohydrates could selectively fortify beneficial gut bacteria like bifidobacteria. I suggested we should give this concept a name and we agreed to think about possibilities. I went home and started drafting the review. About two hours later I sent it to Marcel, who then turned my words into something more resembling science and he drafted the figures (one of which was in color—unheard of back then!). We finished in it a few days.

"prebiotics have safely helped to improve the gut health of people and animals"

Then the argument started about the concept name. I favored "parabiotics." This was driven by one fact alone. At the time, *MASH* (Mobile Army Surgical Hospital) was a popular comedy program in the UK. It was an American series, set in the Korean War, featuring paramedics, and these were people who helped medics. So, a "parabiotic" would help biotics—right? Wrong, according to Marcel, who instead proposed "prebiotics." We went with that and called the paper "Dietary Modulation of the Human Colonic Microbiota—Introducing the Concept of Prebiotics." It was published in the *Journal of Nutrition* in 1995.

Little did we know the impact that a review that took only a few hours to write would have. I can't comment on the quality because I have not read it since publication (in common with the rest of my embarrassing writing). Marcel did all the good bits in there anyway.

25 years later, the following still amazes me:

- The prebiotic concept is now the subject of several conferences, meetings, and workshops each year.

- New prebiotic dietary products have arisen, as well as variations of existing ones.

- Prebiotics is a term often used by consumers, which is unusual in the field of microbiology, which is plagued by jargon.

- I've seen 'prebiotics' written on numerous product boxes and even heard it on many TV and radio adverts.

- In 2018, Yeung et al. (*Food Chemistry* 269, 455–465) reported that our review was the most highly cited of any functional foods papers ever published, with 3,797 citations at that time. This is madness!

- There are now 3,000–5,000 research papers on prebiotics (depending on which website you look at).

- Prebiotics have a value of $3–16 billion worldwide (depending on which website you look at).

- There is a predicted 11–15% economic and scientific rise in this subject over the next five years (depending on which website you look at).

I've always known that the most productive research originates in a bar...but all of this pales into insignificance compared to the instances where prebiotics have safely helped to improve the gut health of people and animals, as they were intended to.

P.S. One common comment I get is, "You should have called it something else." My usual reply is, "Such as?" That usually elicits silence—but I still think I was right with "parabiotics"...

P.P.S. It is not all plain sailing: in 2004 the *Journal of Nutrition* rejected our follow-up review updating the concept."

What a guy!

We can't wait to see where he takes the research next.

chapter ④

bullsh*t
bin

(sorry mum!)

And now we're on to the *how*. As we're walking out of the school playground and back home to digest all the science we've learned, have a wee walk past our new bin, it's like the official charts top rundown of all the nutrition nonsense we see and hear. So, turn the radio up and tune in to the bullsh*t bin!

Skinny teas

These are teas (yes, tea bags!) that claim to make you lose weight but are really just expensive flavored water. Some contain caffeine, a diuretic, and senna, a laxative—peeing and pooing more may make you feel lighter but these ingredients can cause other issues and come with side effects like bloating, cramps, and diarrhea. Long-term use of senna can also impair how well your gut works, meaning you become dependent on it to make you go. Not cool. Not cool at all.

Cleanses, detoxes, lemon water, and celery juice

Aside from lemon potentially eroding the enamel from your teeth, there is no evidence that foods or drinks detox your body—your body is detoxifying all the time. Providing you have a working gut, kidneys, and liver, your body has all it needs to detox.

"Superfoods"

We think most foods have the power to be "superfoods." Whether they provide you with gut-loving fiber or polyphenols (see page 99), or the taste of a cinnamon bun takes you back to being a carefree 10-year-old, the physical and mental benefits food can provide mean there are no true superfoods. Save your cash, avoid expensive powders and exotic ingredients, and instead eat as many plant foods (and colors) as possible and you won't go far wrong.

bullsh*t bin

Clean eating

We hope the fad of #cleaneating passes. A gut-loving diet is one that contains an abundance of plant foods, but that doesn't mean you can't enjoy other ingredients that bring you pleasure as part of a balanced diet. Clean eating has also seen the rise of disordered eating, which has very serious consequences on your gut indeed.

"No carbs before Marbs"

If you want to starve your hardworking microbes of fuel to do their thing, don't starve them of carbs. Whole grains in particular are a really important source of fiber to fuel your microbes, so if you suddenly cut them out, you are going to negatively affect your gut microbes.

Fad diets

They are called fads for a reason. Do your research and speak to a professional if you want to change your diet.

Magic pill

No matter how many supplements you pop, there is no magic pill for a happy gut. We are all individual, as are our guts.

"Fruit is the Devil"

Fruit gets a bad rap and it shouldn't. Fruit contains important fiber, nutrients, and polyphenols for your gut. Try to have it in its whole form, rather than juiced, to get maximum benefits.

The bottom line

Food provides not only nourishment but also enjoyment. Be informed and make educated decisions (that means not treating social media like an encyclopedia of wellness) when thinking about your gut. If common sense tells you it's too good to be true, that's because it is!

myth-busting toolkit

After reading their book *Is Butter a Carb?*, we knew exactly the two gals to arm you with your Inspector-Gadget-style myth-busting toolkit to detect all the nutri-nonsense crimes out there. Enter Helen West and Rosie Saunt of The Rooted Project (slow-mo Charlie's-Angels-style walk)...

"Nutrition myths are everywhere. From claims about the toxicity of whole food groups (e.g. carbs kill!) to the reductive focus on individual nutrients as a cure-all, nutrition myths are used to sell us a vast array of food products, diets, and lifestyles, often with flashy advertising and strong, clear-cut messages. On the surface, these messages are appealing. They simplify the complex area of diet and health and neatly commodify it into something we can purchase and control. Got cancer? Take these vitamins. Feel tired? Follow this detox. Want abs? Buy these appetite suppressants. You get the picture. Bold nutrition claims speak to our (very human) desire for simplicity; they draw on our anxieties about our bodies and health and sell us straightforward solutions and quick fixes. Even knowing this, there are so many people out there exploiting the knowledge gaps and uncertainties in this complex field, that sorting the truths from the half-truths and downright nutri-nonsense can feel like a dizzying task. But you don't need to be an expert in nutrition to navigate the myths. There are often telltale signs that a claim or product may not be quite as it seems.

So, here are our **top ten** tips for spotting nutrition nonsense:

1 A focus on single foods or nutrients
Bold or scary claims about single foods or nutrients causing disease are a sign you should be wary. Health is complex, so these claims are likely to be untrue or, at the very least, overblown.

2 Offers a simple fix for a complex problem
If something sounds too good to be true (e.g. it is offering a single, simple solution, like a vitamin pill, for a complex health problem), then it probably is.

3 Uses 'hot bodies' or celebs to bolster claims
"Eat like me, look like me" is a common unspoken marketing trick for many food products. Are they linking their product with a look? Or claiming a food changes your appearance?

4 Proof via anecdotes
People's experiences and stories are valid and important, but we can't use them as 'proof' that something will work in the same way for us.

5 Promotes one way to eat as the right way
Have they come up with THE way to eat for good health? Healthy eating can be reached in many different ways, so if they are promoting their diet as the one and only road to "healthy," be critical.

Is it selling you a "detox"?

Are they promoting the idea that our environment is toxic and that we need specific products to detox our bodies? Eating well supports good health, but special products aren't needed to improve our body's capacity to detox itself.

A focus on "natural" as best

Saying something is "natural" makes it sound like it is wholesome or pure and, most importantly, "safe." It's comforting and appealing. But "natural" isn't synonymous with any of these things. Many dangerous things are 'natural'—like cyanide.

Is it actually an advert?

Check that the article you are reading is actually an article. Paid promotions for nutrition products like diets or supplements can be dressed up as impartial communications.

Promotes a diet as the sole alternative to proven medical therapies

Any products or services which advise that you should reject and avoid your doctor in favor of their services should be treated with caution.

Uses sensationalist language like "toxic" or "poisonous"

If a communication is sensationalist and using language that invokes fear, then view claims with scepticism."

You're now ready. Go, go gadgets GO.

chapter 5

what can you do?

As we said, we've got your back. Our top tips to support your gut mean making use of the bullsh*t bin (see page 77) and focusing on a few simple things. We have so many tools in our armoury, starting off with the gnashers we were made with, plus a dash of mindfulness...

chew, chew, and chew again

We know this sounds EMBARRASSINGLY SIMPLE but, "Do you chew enough?" As we learned back in biology class (see page 14), digestion doesn't begin in your stomach; it begins before you even put food in your mouth. We've all experienced the mouthwatering sensation when it's just our eyeballs having the feast, and this is enough to kick off the digestive process. It's natural that we're all busy and guilty of ramming a sandwich down our necks as we hotfoot it across the city. But even though everyone else in your train carriage knows about your tuna-fish sub, your gut doesn't, and we need to warn it... "Ready the troops, tastiness incoming!"

You should be trying to chew your food 20–30 times (we know, seems a lot!) before you swallow to make sure it is properly broken down to make less work for the rest of the orchestra waiting below (see pages 14–19).

Some smart aleck once asked us, "What about soup?" Don't be that guy and use your judgement with soup!

We're naturally food inhalers, so we've started using mini sand timers to train ourselves not to finish our meal until the last grain of sand has dropped–very *Crystal Maze* and a good game for kids too.

mindful eating

So, some of the time it's not about WHAT we eat but HOW we eat. For this we'd like you to imagine you're on a date...with yourself.

- Look forward to your food as you would a hot date. Don't eat foods because they are "good" for you but because they are nourishing, and you enjoy them—kinda like picking a dating partner.

- You should always focus on the person you're dining with, so sit down to eat and prepare for digestion. Turn off distractions and focus on your meal/date.

- Rest your knife and fork between mouthfuls and chew your food. Savor the flavors before swallowing. Imagine you're having a conversation and having to leave gaps to chat/boast/giggle/run for the hills. As for date chat suggestions, here's a little starter for 10 to ask yourself before/during eating:

  - Are you responding to an emotional want, thirst, or your body's needs?

  - Is your stomach gurgling? Do you feel hungry or do your energy levels feel low?

  - Are you reaching for food due to a stressful event?

  - Have your food choices been impacted by your emotions? Could you eat nourishing foods to support how you feel instead?

  - Are you focused on your food or is your mind distracted?

  - How does your food taste, smell, and feel?

  - Are you full? Check in with yourself during your meal.

to fast or not to fast?

You've probably heard a lot about fasting, 5:2, 12-hour, 16:8. But what is best for your gut microbes? The truth is, we don't really know, and it varies greatly from person to person, depending on your health and stage of life. Some research shows that giving your gut a break overnight can help your gut microbes to flourish, but whether whole days of food restriction benefits them, regardless of any other purported benefits, the jury is out. Do what you find works for you but pay attention to your body and no one else's... you do you.

variety really is the spice of gut life

So, we've got a whole other family to feed in our guts, and different microbes thrive on different foods (just like our kids!), so variety really is key to making sure you keep them all happy. We all tend to reach for the same fruit and vegetables in the supermarket as we know what to do with them, but the recommendation is to try to eat thirty different types of plant-based foods a week (this includes whole grains, nuts, and seeds). This may seem like quite a high number to hit, but some of our top hacks are:

- Nut and seed mixes.

- Bags of different types of lettuce.

- Stir-fries where you can throw in bits of leftovers and ingredients you need to use up.

- If you don't know what to do with a particular vegetable, extra virgin olive oil, especially the higher-quality types (also high in polyphenols; see page 99), salt, and the oven are your friends for lots of roasted veggies.

  Head to **www.thegutstuff.com** for some handy "gut health at home" resources to help you chart this.

TAKEAWAY
Switch your mindset to ADDING foods, not restricting them.

why is fiber so important?

If we don't get enough fiber in our diets it can result in constipation, decreased diversity of beneficial gut bacteria, and an increased risk of developing certain health conditions. What is fiber? Fiber is a type of carbohydrate either naturally derived from plants or extracted and added into product as isolated fiber. Unlike some carbohydrates, fiber cannot be digested in the small intestine and so passes through to the large intestine where the magic happens. It is an essential nutrient for your gut and allows your gut microbes to thrive. Fiber is food for your gut microbes; they like to ferment it and, in doing so, produce short chain fatty acids. Fiber increases the bulk and softness of your stool and increases gut transit time.

People who enjoy diets rich in fiber (25–29g) experience a 15–30% decrease in cardiovascular-related mortality, incidence of coronary heart disease, stroke, type 2 diabetes, and colorectal cancer, in comparison to low-fiber consumers. The majority of us are not eating enough fiber. We need to consume 30g (1oz) fiber per day as part of a healthy diet (SACN, 2015), which can be a hard number to hit when you think that one medium apple contains approximately 2.1g fiber— that's a lot of apples!

We've talked a lot about plant-based variety throughout this book, but variety of fiber is just as important.

There are many different types of fiber: you'll often hear the term "soluble and insoluble"—this just refers to whether it can be dissolved or not in water. Splitting up fiber in this way doesn't really tell us much about the effect it has on the body, and your gut will react differently depending on the types of fiber and proportions found in any given food.

The key is to get a variety of fibers from different sources. Here's a deeper delve into the key types of fiber (essentially they just have a different chemical structure) and where to find them, which Mother Nature helpfully tends to package up for us to enjoy.

*A word on resistant starch: this is considered a type of fiber, as it cannot be digested in the small intestine. It passes through to your large intestine, where it works like soluble fiber and is fermented by your bacteria, producing short chain fatty acids. You can find it in bananas, potatoes (cooled), pulses, seeds, and grains.

how to get more fiber into your diet?

Our top tips for increasing fiber:

1. Follow our simple swaps (see page 93).

2. Add extra veggies/pulses to ready-made food.

3. Make plants the star of the show. Increase your portions of vegetables and fruit (in that order). While guidelines state we should be aiming for five portions of vegetables and fruit a day, new research shows we need closer to ten, so aim for five portions as a minimum!

4. Choose whole vegetables and fruits over juice.

5. Know your numbers—check out our fiber chart (overleaf).

6. Keep skin on veggies and fruit (where lots of the important fiber is).

7. Include whole grains.

	Cellulose (found in the cells of plant walls)	Hemicelulose (like cellulose but simpler in structure)	Lignans (polyphenols found in plants)	Beta-glucans
Whole grains	☺	☺		
Vegetables	☺	☺		
Fruits	☺	☺		
Seeds	☺	☺	☺	
Nuts	☺	☺		
Bran	☺			
Seaweed				
Legumes/pulses (e.g. peas, beans, chickpeas, lima beans)	☺	☺		
Woody parts of vegetables (e.g. celery)	☺		☺	
Oats/barley	☺			☺
Potatoes				
Onions, garlic	☺			
Endive	☺			
Jerusalem artichoke	☺			

Pectins	Gums and mucilages (used as gelling agents, thickeners, stabilizers, and emulsifying agents)	Resistant starch*	Oligosaccharides
	☺		
☺			
☺			
☺	☺		
☺			
	☺		
☺			
☺		☺	
			☺
			☺
			☺

☺ = decent amounts found in these foods

*If you think your diet is low in fiber, build up your fiber intake slowly: Increase the amount you consume by around 5g a day for a week to limit adverse effects. Some medical conditions mean that increased fiber intake isn't advisable.

TAKEAWAY

If you're upping your fiber, make sure you drink enough water. How much water you drink is going to vary—key factors will include your size, how active you are, and the climate you live in. A great way of seeing how hydrated you are is to check the color of your pee—if it's dark and strong, drink up.

dark chocolate has 3-5g of fiber per 30g (1oz) portion

easy swaps to help you get more fiber in your diet

Swap...	For this	Approximate fiber per portion
White pasta	Wholegrain pasta (80g/2¾oz, dry) Lentil pasta (80g, dry)	4g 4.72g
White bread/pita	Brown pita (1 pita) Rye bread (25g/1oz) Sweet potato (1 medium/160g/5¾oz)	2.6g 10.1g 4.8g
White rice	Brown rice (125g/4½oz, cooked)	3g
Snacks, such as chocolate, cookies, coated rice cakes	Pear and 1 tsp almond butter Banana or apple Orange Hummus (25g/1oz) and carrot batons (1 large carrot/2 medium)	3.5g (pear), 0.5g (almond butter) 2g each 3.1g 1.8g (hummus) 1.9g (carrots)
Cornflakes	Oatmeal (40g/1½oz), sunflower seeds (25g/1oz) and berries (80g/2¾oz)	3.3g (oats) 2.2g (seeds) 1.68g (berries)
Croutons in a salad	Lentils (80g/2¾oz) Chickpeas (80g/2¾oz)	3.68g 6.08g
Fries	Broccoli (100g/3½oz) Carrots (100g/3½oz)	2.6g 2.8g

gut involved in fermented food

We've learned the science and know that bacteria can be our pals, so now is the time to think about adding fermented foods to your diet. Fermented foods contain live bacteria (if they've not been pasteurized or heated) and by-products of what the bacteria make, which may support your gut. Fermenting has been around for centuries and there's plenty to choose from:

Sourdough

Consider this your entry-level fermented food. There's a surprising body of evidence all around our favorite loaf—sourdough. What makes it so special?

- It makes gliadin and glutenin protein more digestive.

- *Lactobacillus* (the bacteria in the "mother" or "starter") produces beneficial by-products to keep your gut happy.

- Sourdough provides minerals iron, calcium, magnesium, and zinc, and the fermentation process helps make these more available.

What makes it sour? Bacteria and yeast produce acetic and lactic acid, giving it a twang.

Sauerkraut

You don't need to head to an expensive health store, it is as simple as some cabbage, a bit of elbow grease and salt. So, roll up your sleeves and find out how. It's great on top of stews and salad—just don't heat it, as it kills off the bacteria.

Kimchi

A Korean side dish of salted, fermented vegetables. Simple but delicious and gives a kick to any meal!

Miso

A Japanese paste made from soybeans to add that umami flavor to food (see, this fermenting stuff really isn't anything new). We love creating dressings with miso paste, kefir, and sesame oil—drizzle on salad or roasted veggies.

Kefir

Fermented milk or water using kefir grains to produce a sour-tasting drink, which can be flavored to suit your taste. Make it yourself—or most supermarkets now sell it.

It's good to learn how to practically introduce these foods to your diet, so here's a little table to show how you can use kefir like you would yogurt:

sweet	savory
Poured on a bowl of oatmeal or in overnight oats	Tzatziki with grated cucumber, a squeeze of lemon, and mint
With stewed apples or berries	Tartar sauce with capers, dill, and shallots
For breakfast with berries and nuts	With lemon and any herbs to make a salad dressing

krauting about

how to make sauerkraut in 6 easy steps!

equipment

large bowl

1 litre glass jar

scales

sharp knife

ingredients

1 head of green or red cabbage (or both)

2% of cabbage weight in salt

1 tbsp fennel seeds (optional)

method

1. Wash your hands and sterilize your equipment.

2. Chop cabbage finely (save core).

3. Weigh cabbage and calculate 2%. Weigh out salt.

4. Massage cabbage, salt & fennel seeds together for 10 minutes. Set aside for 30 mins.

5. Push mix into jar & all the juices. Push core down to submerge cabbage.

6. Store at room temp & burp daily. Test taste at 15 days. Leave for a max of 21 days, then refrigerate.

Live yogurt

Yogurt can be a rich source of beneficial bacteria, which support a happy and healthy gut, but it really depends on what type of yogurt. While bacteria is required to ferment milk (including non-dairy) into yogurt, the manufacturing process, such as heating, that some yogurts go through can kill off the bacteria. Choose an unflavored option and look for "live cultures" or "live active bio cultures" on the label, which means that the bacteria have been added back in or haven't been killed off in the manufacturing process. Different yogurts contain different live cultures, so read the label and mix it up to make sure you are getting plenty of diversity. For example, one brand of Greek yogurt contains the live cultures *Bifidobacterium*, *Lactobacillus bulgaricus*, and *Streptococcus*, whereas another contains *Lactobacillus bulgaricus*, *Streptococcus thermophilus*, *Lactobacillus acidophilus*, *Bifidus*, and *Lactobacillus casei*. Diversity is key for gut health!

Kombucha

Fizzy tea in its finest fermented form. We love kombucha instead of a G&T or as the mixer with gin. Homemade kombucha tends to be a bit punchier on the taste buds, and the store-bought ones are on a sliding scale in terms of efficacy and "traditional fermenting methods." Either way, you'll soon find the right ones to suit your tastes.

As with everything, start slow and build up, otherwise you may find you get some unwanted side effects! A little goes a long way. Fermenting yourself is fun; if you don't want to get your kids or flatmates a pet, try keeping some of these things alive instead—there's lots of how-tos and troubleshooting on our website.

Now you've got your probiotics covered, think about upping your intake of foods high in prebiotics: onions, garlic, leeks, endive, underripe bananas, asparagus, artichokes, olives, plums, and apples, plus grains like bran and nuts such as almonds.

*If you are pregnant, breastfeeding, immunocompromised, or have high histamine levels, fermented foods are not for you—seek advice from your medical professional.

up your polyphenols!

polyphenols is one of those buzzwords that is thrown around, but what does it actually mean?

Polyphenols are natural chemicals found in some plant-based foods with antioxidant and disease-fighting/preventing properties. If something has antioxidant properties, it has the power to combat the negative effects of free radicals, which can cause stress-related cell damage to the body, including aging. Imagine the polyphenols as mini superheroes in your body, fighting and sweeping up the free radicals that your body naturally produces and you are exposed to as part of day-to-day life. You can find polyphenols in all plants, but some contain more than others. Try to eat a rainbow to get a range of different powerful polyphenols in your diet. Foods high in polyphenols include berries, brightly colored and dark leafy vegetables, green and black tea, filtered coffee (yes!), red wine (yes!), and bittersweet chocolate (at least 75% cocoa solids—yes!).

TAKEAWAY
A lot of the ferments you buy in grocery stores will be pasteurized and some, like yogurt, will have bacteria added back in. Always check the back of the label.

fermented foods: is there evidence?

Miguel Toribio-Mateas is a clinical neuroscientist by day and (like us!) a DJ by night. He's the wackiest scientist we know (in the most brilliant way!) and has a wicked sense of humour, plus an extensive background in nutrition practice and research. As part of his visiting research fellowship in microbiome and mental health at the School of Applied Sciences, London South Bank University, he is studying fermented foods, so is PERFECT to tackle this subject...

Miguel Toribio-Mateas

"Conducting research on any kind of food, especially temperamental live foods, can be tricky because of the amount of confounders (yep, first we heard of that word too, but all it means is 'other factors that could influence a result'—very important in clinical trials and how we interpret the evidence) that can affect the outcome of that inquiry. Imagine taking part in a research study on the effects of 'food X' on any particular symptom, let's say bloating. As a researcher I need to be aware of these other factors:

- What about the effects of how hydrated you were during the study, or whether you smoked or drank alcohol?

- What about the effect of how well and how long you slept?

- And what about your physical activity level/exercise?

- Ultimately, was food X definitively responsible for the change in your symptoms? And what potential weight did other foods you ate alongside food X carry?

Next, comes the complication with what I am measuring. When it comes to gut health and its effect on the health of the rest of your body (physical and mental), it is difficult to find one indicator as an absolute marker of health. Clinicians use these indicators (called biomarkers) when diagnosing a condition. One such example is calprotectin, a molecule produced by immune cells in the gut—high levels are found in people with inflammatory bowel disease (IBD), Crohn's disease, or ulcerative colitis. Another issue is how the results are collected. Self-reported symptoms are often reported using a questionnaire to help the researcher understand the severity of the patient's bloating, cramping, and discomfort. But as symptoms are experienced differently by different individuals, this is a very subjective measure and so not necessarily the most reliable.

When researchers run drug clinical trials, there's a typical series of phases the drugs have to pass through before and after they are approved so that they can be prescribed and marketed for a specific use. This process isn't relevant for most food products, but fermented foods are a little different. Why, you may be asking? Well, fermented foods like kefir, sauerkraut, kimchi, miso, and kombucha, etc. have been used traditionally in different cultures for centuries, but now researchers want to assess their value as foods with a specific function (beyond nutrients), or as 'functional foods.' If these functions are backed by research and approved by the relevant authorities, it means health claims about their therapeutic or medicinal properties can be made about a specific fermented food. Without rigorous research, you cannot make such claims. So even if clinical research on food is always going to be a little 'funky,' it is still very much needed.

In my research role as Lead Neuroscientist at the Bowels and Brains Lab at London South Bank University, I've been assessing what kefir, sauerkraut, kimchi, and fermented vegetable juice do to the gut and also to the brain. To figure out the effects of these foods on the gut, researchers have looked at changes in the type and amount of bacteria, such as *Lactobacilli* and *Bifidobacteria*. For the brain, we've used psychometric tests to look at how the brain changes (or not) in response to these types of foods. These two types of results (the effect on gut microbes and the brain) are then fed into a statistical analysis phase to identify patterns between the food and changes to gut symptoms, such as bloating or stool consistency, but also to mood, attention/focus, or memory.

Trials cost a lot of money (sometimes millions) and can take a long time to complete, which means thousands of man/woman hours and a lot of stress in the lab, chasing participants, analyzing the data, and writing up the results, aiming to get them published after an exhausting peer review. (What's a peer review? To be published in a reputable paper, a research study must be peer reviewed—this means it gets looked over by other researchers to validate the study for publication. It is a form of self-regulation and ensures a high standard in the publication of research studies.)

Despite all of these hurdles, this is an enormously exciting time to be involved in clinical research on the effects of fermented foods on health, and it feels hugely enriching to be able to contribute to people's health by providing them with scientific evidence on how foods we've been eating for centuries work by interacting with their gut bugs."

larder

Having a good sort out of your larder and streamlining what you have, not only helps you keep track of what you have but if you can see what you have, you are more likely to use it. This doesn't have to be expensive or require a foraging trip to find exotic ingredients.

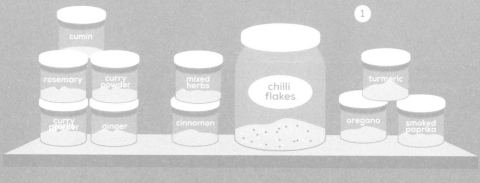

1

cumin

rosemary

curry powder

mixed herbs

chilli flakes

turmeric

curry powder

ginger

cinnamon

oregano

smoked paprika

2

oats

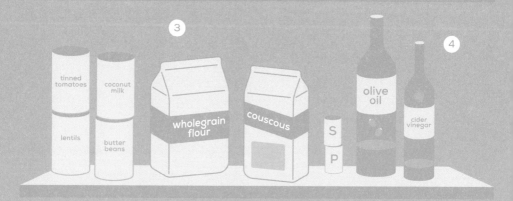

3

4

tinned tomatoes

coconut milk

wholegrain flour

couscous

olive oil

cider vinegar

lentils

butter beans

S

P

herbs and spices—These can make a huge difference to some otherwise rather bland kitchen-cupboard essentials and all add variety for your gut microbes. Our staples are:

Turmeric: great for adding color and an earthy flavor to curries

Cinnamon: add to smoothies, stewed fruits, banana bread, and oatmeal for a warming feel

Smoked paprika: great for Mexican or Spanish flavors; we like ours in baked eggs and on sweet potato fries and hummus

Oregano: gives instant Mediterranean flavor. Mix it into salad dressings or use on roast veggies

Chilli flakes: we love chili on our hummus or when we want to take a dish up a notch

Ground ginger: add to curries and stir-fries for extra flavor, and no grating needed!

Garlic powder: easy to add to dishes and saves the fuss of crushing it (and no smelly hands—win!)

dry goods—A difficult one to condense into a shortlist, but a few staples will help you to make a meal using a wide variety of ingredients.

Dried pulses: a little goes a long way, but you will need to soak them in water overnight; our favorites are beans, peas, chickpeas, and lentils.

Quinoa

Pasta: wholegrain, lentil, and chickpea

Soba noodles

Wholegrain couscous: makes a super speedy lunch

Pearl barley: good to add to soups or to use instead of other grains

Oats: great for making overnight oats, oat bars, oatmeal, or savory oats

Wholegrain flour

Nuts and seeds

cans—Great for convenience and boosting nutritional value of meals.

Chopped tomatoes, Coconut milk, Beans and pulses

oil & vinegar

Good-quality extra virgin olive oil: packed full of polyphenols

Apple cider vinegar (with the mother)

fridge/ freezer

Having a system for your fridge and freezer can make it a lot easier to use and if you can see what you've stored, it will encourage you to use. We use glass jars (and recycle the ones food came in) and glass boxes to help us see what we've got.

① top shelf—stuff you've made:

Hummus (homemade or bought)

Stewed fruits (we try whatever we can find)

Pre-cooked vegetables (always do a bit extra at dinner for the next day)

Pesto (we use whatever nuts we have to hand and then something green—basil, spinach or rocket)

② middle shelf—dairy and fermented foods:

Artisan cheese—these are the ones that typically contain plenty of bacteria (if unpasteurised)

Live yogurt (always read the label)

Kefir (milk or water)

Kombucha

Ferments (we try cabbage or root veg)

③ bottom shelf:

Meats

Fish

Vegetables / salad / fruit box—just aim for as much variety and as many colours as possible

④ freezer:

Diced apple, berries of any kind, sliced banana out of its skin

Vegetable and fruit lolly pops—made using a mix of both and kefir or live yogurt

Vegetables—spinach and mixed veg

Pasta sauces (we freeze portions in a large ice cube trays so it is super easy to lift out)

Muffins, energy balls (freeze separately—that way you just lift out what you need)

Pre-sliced sourdough—super easy to lift out a slice when you need and prevents waste

Batch cooking—curries, soups, lasagne, fish pie, bean burgers, banana pancakes

simple swaps

Our aim has always been to make gut health accessible for everyone, which is why we've got some simple swaps for you to implement a bit at a time to help you on your gut health journey. Take it steady and keep it simple—you don't have to go from zero to hero as soon as you put this book down.

1. swap white for brown

a. Change white pasta for wholegrain—white pasta (2g fiber) vs wholegrain (7.5g fiber).

b. Instead of white rice, try brown, red, or black rice—white rice (1g fiber) vs brown (4g fiber).

c. Swap low-fiber cereal for a fiber-dense variety, such as bran or oat-based cereals—cornflakes (0.8g fiber) vs bran flakes (5.4g fiber).

d. Try wholegrain sourdough or rye bread—1 slice of white bread (1g fiber) vs 1 slice of seeded wholegrain (5g fiber).

If you aren't a fan of wholegrain versions of your white favorites, try introducing them into your diet slowly (e.g. use half white/half brown rice) until you get used to the different texture and flavor. Making this simple switch will help you reach 30g fiber a day.

2. vegetable boxes

Instead of opting for your usual online food shop, why don't you give a vegetable box a whirl? Vegetable and fruit boxes often work out to be great value for good-quality produce and a lot of the schemes support local farms. It's also another way of trying to get more plant-based variety over the course of the week as the contents change regularly depending on what's in season. There are lots of great options out there, from odd-shaped veggies to farm-to-home boxes. It's a great way of getting near thirty different plant-based foods a week to support that diverse ecosystem in your gut.

3. make the freezer your friend!

You tend to get more food for your money by buying frozen and you won't be compromising on quality. Spinach, berries, and mixed vegetables are great freezer staples and give you the opportunity to increase variety without food waste.

4. swap regular pasta...

...for chickpea, lentil, or pea pasta to increase the different plant-based foods in your diet.

5. yogurt

Where a recipe calls for yogurt, replace with live plain yogurt or kefir.

6. chocolate

Swap milk chocolate for bittersweet (anything north of 75% cocoa solids) to increase your consumption of gut-loving polyphenols.

7. roasted vegetables

Swap skinless roasted veggies for skin-on where you can; just be sure to give your vegetables a good scrub.

8. condiments

If you are a condiment lover, swap your usual condiments for sauerkraut or kimchi. You'll be amazed how your taste buds change to crave their fermented flavors.

9. soup

Swap your canned soup for miso soup.

10. sodas

Swap for kombucha or, if that's going too far, sparkling water with some fresh herbs or mint to reduce your intake of artificial preservatives and sweeteners.

11. ditch the guilt!

Enjoy your food. Digestion starts before food passes your lips—enjoying and looking forward to the food you eat is just as important as chewing properly. If you are holding your nose and downing a celery juice, you aren't going to digest it properly.

alcohol

Excess alcohol can aggravate your gut, from how well you digest your food to things moving a bit quicker than usual!

Too much alcohol can inhibit the digestive enzymes you produce, which means your food won't be digested and absorbed as well as it could be.

Undigested food makes more work for your gut bacteria and when your bacteria has to work harder, you produce more gas—cue bloating and windiness when you've had a few too many cocktails. There's a lovely bacteria that lives in the mucus lining of your gut—remember *Akkermansia* from the job description on page 23? When you drink a lot of alcohol, the mucus lining can get damaged and this can reduce the amount of Akkermansia as well as other important microbes.

There's only one man we could draft in to answer our many, many questions on alcohol (usually via WhatsApp on hungover Sundays: "How much? What kind? Why is this hangover so bloomin' awful?").

Gautam Mehta is a liver specialist and honorary consultant at University College London, with many incredible research projects on the go, but he took a break to answer these (very important!) questions for us all...

Gautam Mehta

is red wine REALLY good for you?

"Research shows there is an association between moderate red wine consumption (see opposite for what 'moderation' means) and increased diversity of gut microbiota, which in turn has been associated with a healthier gut and possible benefits for cardiovascular and other metabolic diseases. Before you grab a bottle of red, the key word in this last sentence is 'association'—this doesn't prove cause and effect. The study in question looked at a snapshot of the gut microbiome and compared this with alcohol and dietary information in the study population, rather than conducting a prospective intervention study with a control group. To me and you, that means people self-reported consumption and diet instead of being prescribed a diet and specific units of alcohol to drink (and people tend to underreport on these things). Your microbiome has the ability to change very quickly, so a one-off stool sample doesn't always give a clear picture. Nevertheless, the effect seen was much greater for red wine than for white wine or other alcoholic drinks, which may be because of the greater concentration of polyphenols in red wine. Polyphenols seem to inhibit the growth of certain gut bacteria and promote the growth of others and are associated with beneficial health effects. The other important fact to note, though, is that polyphenols are also found in high levels in fruits, vegetables, coffee, and tea, so red wine isn't a necessary dietary supplement to get more of them in your diet."

is a big session really worse than a couple of glasses at home?

"The other important factor to consider is the amount of alcohol consumed in one sitting. For red wine, only a couple of glasses a week was required to see the benefits in gut bacteria described opposite. In fact, there is good evidence that binge drinking—in this case a binge is more than 6 units for women or 8 units for men (a large glass of wine has 3 units, a shot has 1 unit)—leads to increased intestinal permeability (or "leaky gut") and increased inflammatory proteins in the blood (head over to page 55 for more on this). This is probably due to one of the metabolites of alcohol called acetaldehyde having a direct effect on the gut lining. Over time, it's possible that repeated inflammatory "hits" may lead to chronic inflammation, which is linked to a number of chronic diseases, including mental health conditions (see page 55). It's difficult to say how long it takes to recover fully from a binge, but our experiments with the BBC's *Horizon* ("Is Binge Drinking Really That Bad?" in 2015) suggest it takes at least a week, probably two. It's also important to say that regular, low-level drinking also leads to many other negative health consequences, such as an increased risk of cancer, but does not have the same effect on gut leakiness."

what is it that gives us the hangover?

"As mentioned above, acetaldehyde is one of the intermediate compounds generated in the metabolism of alcohol. Usually, acetaldehyde is broken down in the liver by an enzyme called aldehyde dehydrogenase. But this enzyme has a finite capacity and if it becomes saturated by excess alcohol during a binge, then acetaldehyde can start to accumulate. Acetaldehyde can cause nausea and flushing, contributing to some of the symptoms of a hangover (the other symptoms are probably because of leaky gut). So, one way to stop acetaldehyde accumulation is to slow down your drinking, helping to ensure that the enzymes metabolizing the alcohol can keep pace."

orthorexia and the effect on the gut

this next topic is a tricky one and we knew

there was only one woman for the job...

Renee McGregor is a measured, brilliant sports and eating disorder specialist dietician with twenty years' experience working in clinical and performance nutrition. When not inspiring others with her incredible work, Renee can be found running in the mountains and chasing the trails, most likely training for a crazy ultra-marathon.

Renee McGregor

"Orthorexia is an eating disorder defined by "the obsession with eating correctly or purely." It has often been associated with the trend to #eatclean. At the time of writing, there is no specific diagnostic criteria due to the complex way it presents in individuals. One of the key difficulties is that so many of the "wellness" trends of modern life act as a mask to disguise what's really going on. Individuals with orthorexia create strict food and/or exercise rules based on a belief that it will lead to a more healthful life. However, this becomes so obsessive, that any deviation from the rules results in high levels of anxiety, making it impossible for the individual to take part in daily life. As with all eating disorders, in reality, the issue is never food; this is just the medium by which sufferers choose to express their discomfort. In fact, in most cases, an eating disorder acts as a way to deny and restrict difficult emotions that people do not want to experience. Through their rituals around food and exercise, they aim to contain these difficult emotions, creating an illusion that they are in control and maintaining order.

The food rules created in orthorexia usually result in restricting the body's intake of specific food groups and/or overall energy. This means that there is less energy available to the body for the work it needs to do; when we talk about "work," we mean the biological processes that go on inside us to keep us alive, as well as any physical activity we do, voluntary and involuntary. When we are in a state of low energy availability, the body will go into preservation mode and down-regulate biological processes, including digestion.

Simply, there is not enough energy for digestion to work efficiently. Transit time through the gut slows, which is known as gastroparesis, and the symptoms experienced can often be mistaken for IBS. This is often further exacerbated by the fact that, usually, the individual has a diet very high in fruit and vegetables in order to displace the essential energy they need, which overloads the system further. However, a full history and assessment of the individual should be done, as putting someone who is already restricting their diet on a further restrictive diet, such as the FODMAP (Fermentable Oligo-, Di-, Mono-saccharides and Polyols) diet, can actually make the situation worse as well as feed into their already disordered relationship with food.

If someone is restricting their nutritional intake, this needs to be corrected prior to treating the gut symptoms. Working with a specialist dietitian who can do a full clinical assessment to rule out an underlying restrictive eating disorder as a cause for gut symptoms is critical."

If you think this applies to you or someone you know, speak to your medical professional.

keep a diary

m	t	w	t	f	s	s
				1	2	3	4	5	6

m	t	w	t	f	s	s
		1	2	3	4	5	6
7	8	9	10	11	12	13
14	15	16	17	18	19	20
21	22	23	24	25	26	27
28	29	30	31			

this is not a food diary for calorie counting...

...but one to monitor what you're eating, how you are feeling (mentally and physically), your poos (yes, really!), and how much you are moving, to help you tune in to your body and to spot patterns.

We all live such fast-paced lives that we rarely think about how our bodies feel unless we're ill or hungover. Understanding and listening to your body is a big part of understanding your gut, and often we don't listen to it until there's something wrong. You'll be surprised at how much you notice.

Experiment—try taking things out of your routine then putting them back in and seeing what happens. You may also see some patterns between stress and your digestion.

If you do have gut symptoms, you can take your ready-made diary to a nutritionist, dietitian or doctor (we're sure they'll thank you for it). Head to The Gut Shop at thegutstuff.com

de-stress

stress is often a driver behind many different gut symptoms...

...and so it makes sense to learn how to manage stress and calm down—easier said than done. Doing activities that relax you, such as deep breathing, yoga, and meditation (or even having a laugh with a friend) encourages your body to activate the parasympathetic nervous system (rest and digest) instead of the sympathetic nervous system (fight or flight). Being in fight or flight mode is a good thing, providing it activates at the right time, but not if you are trying to digest food or need to get to sleep. If yoga or meditation ain't your bag, give Dr Rabia's breathing exercise a go.

myth bust

"de-stressing must be about gong baths and candles."

TAKEAWAY
Take three deep breaths before eating, enough to switch to rest-and-digest mode.

breathwork
with dr rabia

Before we'd tried breathwork we were like, "breath-what? We do that anyway, surely?"

Then we did a session with Dr Rabia, a gastroenterologist and a yoga instructor (what a combo!), and we knew we had to share her magic with you...

**"Take a deep inhale...
pause at the top...
exhale out through the mouth."**

"What is the power of conscious breathing? What does it have to do with your digestive health? Our story begins with the close relationship between your nervous system and gut: as we have now understood, the two are married from birth. Modern life favors a swing toward 'fight or flight' mode (the so-called sympathetic drive), which can dictate our eating habits, pain threshold, and anxiety levels. The rhythm of our body in times of stress—even if we are not fully aware of it—often involves shallow and rapid breathing, primarily in the upper chest, and is associated with an increased heart rate. This synchrony of heart rate and breathing is called Respiratory Sinus Arrhythmia (RSA) and is part of the normal, automatic processes governed by our trusty vagus nerve (see page 54). Therefore, this pattern also reflects the tone in our belly, as the vagus nerve communicates the motion within our gut.

But how can we nudge the conversation in the other direction? Put simply, we can influence vagus nerve activity by consciously altering the rate and depth of our breathing. RSA is a signal that appears like waves on the ocean—and you are steering the ship. Interestingly, this surrogate marker of vagal nerve activity is most influenced during the exhalation portion of our breath cycle. Extending your exhale just slightly longer than your inhale and slowing down your rate of breathing provides a kind of massage for the vagus nerve. This can correspond to better synchronization between the brain and gut by enabling us to regulate our stress levels and our perception of symptoms.

If you were guided by my instruction at the start to indulge in a single, deep sigh (feel free to do it again and observe), you will notice that the very act of simply bringing awareness to your breath provides a conscious shift to slow down and check in. We all naturally breathe at different rates, oscillating depending on the circumstance. My first tip is to notice what your breathing is telling you right now. With this awareness, bring your attention and breath down into your belly and relax the muscles of the abdominal wall. We do this by engaging the large, sling-like muscle at the bottom of the ribs: our diaphragm. When the movement of this muscle is out of sync with the muscles in our belly, it can exacerbate bloating and discomfort. So, my second tip is to soften the belly: allow the rise and fall of each breath to occur from this space. In many ways the journey to ease, comfort, and control begins with a single, deep breath."

other sh*t you should know

you may never have considered how you poo, but

toddlers have got it right with this potty business.

Sitting on a Western-style toilet means your puborectalis muscle contracts, which means your rectum is tight, making evacuation of poo difficult, straining and putting pressure on your pelvic floor.

Squatting with your feet on an 8–12-inch-high step relaxes your puborectalis muscle (allowing your rectum to open properly), making it easier for poo to evacuate and preventing straining.

Get yourself a step or a box (or even a pile of books!) to put near your toilet and gut in POOsition!

how do you poo?

puborectalis
muscle contracts
squeezing rectum

increases straining and
pelvic floor pressure

• • •

kinda like
a water
flume with
a kink in it

squatting

eg. feet on step
20–30cm high

protects
nerves

puborectalis
muscle relaxes
allowing rectum
to open

prevents
straining

analrectal
straightened

gut in
POOsition!

chapter 6

i've gutta problem

when to worry about your gut

When we originally set up The Gut Stuff we just wanted to shout from the rooftops about all the cool science around the microbiome we were learning about, but as we delved in, we realized just how many people were living with digestive issues, and, even more worryingly, in silence. Most of us hide behind "the poo taboo" and lose track into adulthood of how fascinated we are by the stuff as kids. With babies (and puppies!), we always use poos as an indication of how they're getting on (we have proper doggy-doo inspections in our garden with our puppy!). We need to reignite this curiosity and perform some close monitoring on ourselves.

In this next section we'll be teaching you how to peek before you flush, what to look out for and how to monitor other symptoms, such as bloating, gas, pain, and all the other uncomfortable stuff, to help you spot the signs that something might be up. One in four of us has a digestive symptom at any one time, and these can be confusing as they often overlap. This section should help tease out the issues and teach you how to tune in.

We knew we'd need a VERY safe pair of hands for this—Sophie Medlin is a registered dietician and co-founder of CityDietitians. Sophie has been a specialist colorectal dietician for over ten years. She works with patients who have conditions such as Crohn's disease, ulcerative colitis, bowel cancer, and diverticular disease to help improve their nutritional status and their experience of eating, so you can see why we had to have her for this chapter. Take it away Soph...

Sophie Medlin

"This is what you need to look out for in case something isn't right. First of all, let's cover what is "normal":

Frequency—Anything from having your bowels open three times per day to three times per week.

Color—Normal, healthy stools are usually a dark brown color like chocolate or horse chestnuts. There are some exceptions to this rule, with certain foods; for example, when you eat beet, sometimes they can turn a funky purple color!

Consistency—A normal stool is between a three and a four on the Bristol Stool Form Scale (see page 128). This means that it holds together like a sausage, but it shouldn't cause pain or require a lot of straining to get it out.

Wind—Having wind is completely normal and the amount will vary from day to day, depending on what you've eaten and how much exercise you've done, for example.

Pain—Occasional, mild stomach ache that passes when your bowels open is not a concern.

Bloating—Mild bloating after eating certain foods or a big meal is common and is not a cause for concern unless it is bothering you or it is particularly frequent.

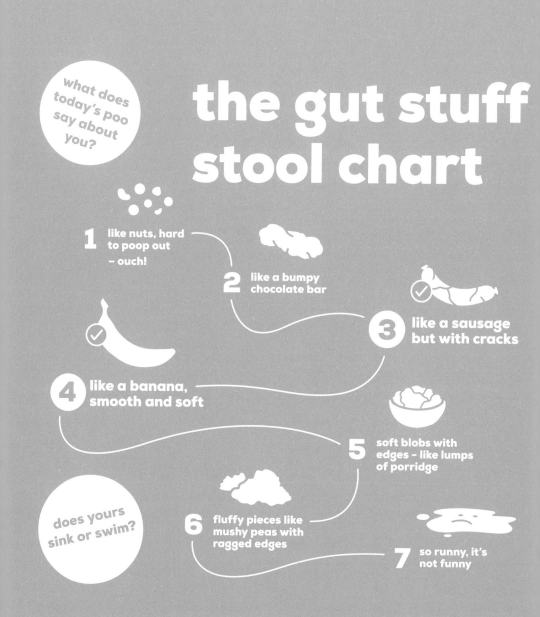

the gut stuff stool chart

what does today's poo say about you?

1 like nuts, hard to poop out – ouch!

2 like a bumpy chocolate bar

3 like a sausage but with cracks

4 like a banana, smooth and soft

5 soft blobs with edges – like lumps of porridge

does yours sink or swim?

6 fluffy pieces like mushy peas with ragged edges

7 so runny, it's not funny

The first step in diagnosing bowel problems is usually a blood or stool test so there is nothing to be nervous about when you're speaking to your doctor for the first time.

Your stools (the technical term for poo) come in all kinds of shades, shapes, and sizes, and sometimes they come more frequently and sometimes not as often. Because we're not very good at talking about our poo, we sometimes worry unnecessarily, and sometimes we don't tell our doctor things that they do need to know.

Your body is really clever at giving you clues when something isn't right. To understand the symptoms, you need to talk to your doctor about them. Remember, it's about YOU, and we're all individual, so it's important to understand there's no right or wrong—it's about tuning in and getting to know your patterns so you can tell when something is up.

things to report!

How often you go

Anywhere between three times a day and once every three days is considered normal.

Changes to your stools

On different days and sometimes at different times of the month (if you have periods, see page 63), your stool might change for a few days. This is completely normal. If you notice a change in your poo that lasts for more than a couple of weeks it is important to tell your doctor. Even if this means that you notice you're suddenly less constipated and you don't know why.

red flags!

If you experience any of these symptoms, you need to speak to your doctor:

Blood in your stools

If there is blood on the toilet paper, also check to see if there is any blood mixed in with the poo. This will help your doctor to know where the blood is coming from. Old blood in your stool looks like black tar, so keep an eye out for that too.

Persistent pain or bloating

This means pain and bloating that doesn't go away when you pass a stool or pass wind, or if you are woken up in the night with pain and bloating. Also, if you notice that you get pain and bloating that builds up and that you have a sudden great urgency to go; then when you do go, you pass hard stools followed by liquid.

Sand- or yellow-colored stools

Stools that are a pale sandy color or are yellow need to be investigated. If you're losing weight as well, make sure you tell your doctor that too.

Poo that is green usually means you've got a stomach bug and generally lasts a few days. Any longer, seek help from a doctor.

Stools that are frothy, foamy or contain mucus

These symptoms can happen after a bug or if you've eaten something that didn't agree with you, but if it doesn't settle down in a few days, speak to your doctor.

Stools that don't flush away

Stools that regularly block the toilet, or that float and take several flushes to get rid of, need to be reported, particularly if you can see undigested food in your stools. You also need to let your doctor know if you can see an oily residue in the toilet after you've been. Usually this will be yellowish in color.

Stools that smell very bad

The question we ask patients is: "Do you think your stools are particularly foul smelling?" If they are regularly smellier than you think they should be but you're already eating a healthy diet, then it is worth getting this checked out.

Other things!

If you notice your poo getting thin and pencil-like or ribbony, then that needs to be reported to your doctor, as does "wet wind," which is the technical term for following through when you think it's just gas! Reporting all these symptoms can help your doctor or dietitian to decide which tests you might need to diagnose your problem. Some of these symptoms suggest a problem with how you are absorbing your food. Some suggest a problem with how your stools are moving along the bowel and some suggest inflammation. Whatever is going on in there, it needs to be investigated and your doctor might not always know the right questions to ask, so make sure you're ready to share the details. Think about keeping a food and symptom diary (see page 117) to share if you find it hard to say the words out loud.

We know this is a lot of information to take in, so Sophie helped us create this to get our heads around it all...

i've gutta
problem

1. Stools that are pale in color, foul smelling, frothy, oily, or difficult to flush away are associated with not absorbing all your food properly. Sometimes this is linked to your gall bladder, bile metabolism, or pancreas.

2. Stools that are black and tar-like are sometimes caused by blood that is coming from something higher up in your gut, such as stomach ulcers.

3. Sometimes, there might be a narrowing in your bowel caused by a scar from surgery, old inflammation, an intermittent collapse, in your bowel or a growth. This can make you feel a build-up of pain and bloating followed by a sudden need to rush to have your bowels open.

4. If there is ulceration or inflammation in your bowel you may have blood or mucus in your stool and you may lose weight.

5. Sometimes you can get pockets in your bowel wall called diverticula. For some people this causes pain, diarrhea, and a fever.

6. Some people don't have very strong muscle contractions to push the stool along the bowel, so the stool stays in the colon for too long, causing constipation. This is called dysmotility.

7. If there is a narrowing in your rectum, which can be caused by scars from childbirth (having a baby), you can have thin, pencil-like stools, or ribbon-like poo.

8. Sometimes, piles cause bleeding when you wipe, or blood to be mixed in with the stools.

IBS: the story of an irritable gut

The thing we get asked about most with The Gut Stuff is IBS. It's complex, it's misunderstood, and it's definitely on the rise. Every time we have a panel event on this topic, Laura Tilt is at the top of our speed-dial list. Laura is a registered dietician and health writer and we've never seen a question on IBS she hasn't been able to answer...

Irritable: abnormally sensitive
Bowel: part of the gut, made up of the small bowel (small intestine) and the large bowel (colon and rectum)
Syndrome: a group of symptoms

The chances are you've heard of irritable bowel syndrome (IBS) or know someone affected by it. IBS is a common digestive condition affecting 1 in 7 adults worldwide. Research shows it's twice as common in women than men (which some scientists think might be related to the impact of female sex hormones on gut function and sensitivity) and is usually diagnosed before the age of 40, although a survey of 2,000 people with IBS conducted by the International Foundation for Gastrointestinal Disorders found that the average amount of time people suffered with symptoms before being diagnosed was 6.6 years.

what exactly is IBS?

IBS is a condition which affects how the gut moves and functions, causing symptoms like pain, bloating, wind, and a change in poo habits—typically constipation, diarrhea, or a mixture of both. We can think of the word "irritable" in IBS as meaning "abnormally sensitive." It's a good description for a gut affected by IBS!

what causes IBS?

We still don't know what causes IBS, but there are a few factors that scientists believe play a role. About 60% of people with IBS have super sensitive nerve endings in their gut, which means they are more likely to feel pain and discomfort in response to normal gas production in the large intestine.

There's also evidence that changes in the balance of gut bacteria, stress, and previous gut infections can contribute to the development of IBS. And 1 in 3 people with IBS have unusually fast or slow gut motility (movement), which affects how often they poo.

can IBS be treated?

IBS is a chronic condition—meaning that it persists long-term. This doesn't mean spending the whole time in pain or discomfort though. IBS symptoms fluctuate, and it's possible to go for months or years with symptoms well controlled.

I appreciate that a diagnosis of IBS can sometimes feel like a nothing-y conclusion—something which is diagnosed after everything else has been ruled out—but I want to emphasize that IBS is a very real condition that can significantly impact daily life. I also want to reassure anyone suffering that there are many effective tools that can help improve symptoms.

what can you do if you think you have IBS?

get a medical diagnosis
The first step is to visit your doctor, as IBS shares symptoms with other conditions like celiac disease, so it's important to get the right diagnosis.

know your IBS
Not all IBS is the same—understanding your symptoms and triggers will help you learn how to manage them. Keeping a food and symptom diary is the first step (see page 117).

begin with the basics
Eating regular meals, reducing your intake of caffeine, alcohol, and spicy foods, and adjusting your fiber consumption are all simple steps that can really make a difference. Learn more at https://www.bda.uk.com/resource/irritable-bowel-syndrome-diet.html

take stress seriously
Research demonstrates a strong link between stress and IBS severity. Put simply, when you feel stressed, your gut feels it too! Although living completely stress-free isn't possible, learning to manage stress and increase your resilience will help. Gentle exercise, breathing practices, and yoga have all been shown to improve symptoms, so it's worth investing time to discover which activities help you to relax and stay calm under pressure.

seek further help
If your symptoms don't improve after trying lifestyle and dietary changes, chat to your doctor about the next steps. You could request to see a dietitian who helps with specialist IBS advice. Approaches like CBT (cognitive behavioral therapy), a type of talking therapy which addresses how thoughts affect behavior, and gut-directed hypnotherapy (which uses guided imagery to help relieve pain) are also available.

a note on fiber

While most of us can benefit from eating more fiber, if you have an existing gut condition like IBS, increasing fiber may aggravate your symptoms. There isn't enough evidence to say what the right amount of fiber is for people with IBS, so experiment with what you can tolerate, and remember that managing your IBS may mean following a different diet to what's recommended for those without a gut condition. As a guide, if you mostly experience loose poos or diarrhea, choosing lower-fiber cereals, reducing fiber from pulses and raw veggies, and avoiding skins, pips, and peels can help. If you mostly experience hard small stools (constipation), increasing fiber from oats, whole grains, flaxseeds, fruits, and vegetables may help. If you are increasing your fiber intake, do so slowly and drink plenty of water too.

chapter 7

the future of science

what does the future hold?

there's so much incredible science emerging

that can really help people with digestive issues...

...for example, fecal transplants, aka "transpoosions" (uh-huh, you heard right!), which are showing very promising results in treating *C. difficile* infections. It always fascinates us when the science starts to translate into tangible advice as well as products and services, such as personalized probiotics, and toilets that can analyze your poo as it drops into the pan. There are many people dedicating their lives to this and our guest contributor in this chapter is one of them.

Step into the tardis, as we time travel with scientist Dr Ruairi Robertson. He has a BSc in Human Nutrition and a PhD in Microbiology and is currently based at The Blizard Institute, Queen Mary University of London. He is also incredibly funny, and we like to sup beers and chat to him for hours to pick his brains. Even though he's young, he's covered so many areas already, so he's the perfect man to show us how the future of gut health science could look...

"Ten years ago, you would struggle to find a photo of homemade sauerkraut or kefir on your social media feed, no-one had heard of the gut–brain axis and certainly no-one was delightedly sending off their faecal samples in the post for analysis. The popularity of gut health in recent years is thanks to the amazing work of scientists around the world who became fascinated by our intestines and their microbes long before it was cool. Thanks to the efforts of these gut scientists, we now know just how important your gut and its microbes are for your health, and words like microbiome and kombucha are slowly falling into our everyday vocabulary. But despite these fascinating discoveries in recent years, there is still so much more to learn about our bodies' most interesting organ and how we can harness it to keep us healthy. So, what's happening next in the labs of these gut scientists that will propel us into the future of gut health?"

home-testing

DNA sequencing refers to the method scientists use to read the genetic code of any living thing. This is currently how scientists find out what microbes are living in your gut. The main reason that we have discovered so much about the gut and its microbes in recent years is because of technological advances in DNA sequencing, which have made it much cheaper than ever before. (In the year 2000, it cost $100,000,000 to sequence the entire human genome, while in 2020 it costs about $1,000.) With these costs declining rapidly, it is very likely that you could sequence your gut microbiome in your own home in the near future. In fact, right now, there are companies working on sequencing machines that are small enough to attach to your phone. It's a real possibility that, in a few years' time, you could test your own gut microbiome after every meal that you eat, or any time you have gut symptoms, to see which microbes are missing. This would probably involve some sort of device that sits in your toilet and connects via bluetooth to an app on your phone to give you real-time updates on your gut microbes.

beyond bacterial catalogs

Most of the successful gut health research in the last ten years has focused on "who's there," meaning that scientists have mainly been interested in listing which microbes are in the guts of people who are healthy or diseased. But each of the trillions of bacteria living inside us have many different skills, which they will use depending on whether they are hungry, tired, or stressed. So, it is now important to focus on "what they're doing." We know that we have slightly more bacterial cells in our body than human cells, but we have a hundred times more bacterial genes than human genes. And genes define what each of these bacteria can do. Think of it like a soccer team: it is helpful to know how many defenders, midfielders, strikers, and goalkeepers you have in the team, but if you don't know who is good at dribbling, who is best at passing, and who is best at shooting, then you can't make your team better. This is the next step for gut scientists: to find out what each of our gut microbes are doing and how this changes with disease.

Scientists are now investigating this using lots of fancy techniques, such as metabolomics, proteomics, and transcriptomics (collectively known as "omics"). In fact, some studies show that by using these 'omics techniques, scientists can tell more about the health of a person by looking at what their gut microbes are doing rather than just what microbes are present in their gut.

personalized treatments

Despite all of the excitement and knowledge we have gained about our gut and its microbes, gut "treatments" remain quite limited. There are hundreds of different probiotic and prebiotic supplements on the shelves, but only a handful of them have good scientific evidence behind them. Even those supported by solid science may work for some people and not for others. This is due to the huge differences in each of our microbiomes, which are as unique as our own fingerprints. Some people's guts may like a certain probiotic/prebiotic, while others may not. That's why personalized gut treatments are a hot topic in the gut science world. Some fascinating studies show that scientists can look at a person's gut microbiome and other markers of health and predict whether their blood sugar will spike after eating a certain food. Using this prediction algorithm, they can design personalized diets for individuals to keep their blood sugar down. This will be the future of gut health. By using fancy AI and machine-learning algorithms, it will soon be possible for scientists, doctors, and nutritionists to design personalized probiotics, prebiotics, diets, or other treatments that will work for your own unique gut.

you've made it to the end!

from fiber and feces to breathwork and brain health, what a journey this book has been.

We hope that not only have you got a grip on some of the science of your body (and around it!) but have also gained some tips on how to arrange your fridge, spot signs and symptoms, and are even inspired to have go at fermenting. Once we all know why, it's so much easier to grasp the how and what, as we have the impetus to make some positive changes, however small and incremental they may be.

We never expected this to be our path in life, but the importance and immediacy of finding out about and looking after our gut health has, quite literally, consumed us. We hope these (exceptionally pink!) pages have shown you even a fraction of our passion for this subject. We've brought together the all-stars from a variety of fields so you can see how widespread (and brilliant) this area of research is.

More than anything, we hope you feel empowered. Everyone has the right to know about their own bodies and, importantly, their guts, and be armed with the tools and confidence to make health and lifestyle decisions for themselves.

If you'd told us, aged 14, we would end up writing a book about gut health, we would have thrown our fries at you and shouted "WITCHCRAFT!" But here we are. Life gets gutsier and more wonderful every day.

Lisa and Alana x

PS: Spread the good gut word and maybe pass this book on to a friend once you're done or visit our website **www.thegutstuff.com**

chapter ⑧

the gut glossary

If you've got to the end of this book and still think, "huh?!", just read this and you'll be right up to speed ;)

bacteria

what is it?

Bacteria are single-celled microorganisms and can be found all over our bodies, including our mouths, skin, and, of course, our gut.

There are so many different species of bacteria in your gut. Think of the different species as having different job titles—they all do different things.

Within a bacteria species (or job title), there are "strains." Strains are a further way of separating out the differences between bacteria belonging to the same species. For example:

Species: *Lactobacillus* (music artist)
Strains: *L. acidophilus* (bassist)
       L. amylovorus (DJ)
       L. casei (singer)
       L. rhamnosus (backing vocalist)

what does it do?

Mostly, lots of great things, but sometimes the wrong type or too many/too few bacteria can cause problems.

The bacteria in your gut make up your microbiota.

The ratio of human to bacterial cells is thought to be 1.3:1. Scientists are continuing to research the true ratio, but it makes up a big proportion of who we are!

bile

what is it?
Bile is a liquid produced in your liver and stored in your gallbladder. Your gut microbes can influence your bile and bile can influence your gut microbiota.

what does it do?
Bile helps us break down fats so we can absorb fat-soluble vitamins, such as vitamins A, D, E, and K. It also gets rid of toxins we no longer need.

chyme

what is it?
Chyme is a cocktail of stomach acid, digestive enzymes, partially digested food, and water.

what does it do?
Chyme allows for further digestion by enzymes and is a carrier of food and enzymes to the small intestine.

dysbiosis

what is it?

Think of your gut microbiota as a country garden—you want a variety of different plants (beneficial microbes) and not too many weeds (fewer beneficial microbes) for a happy and healthy gut. We need our gut microbiota to be diverse and balanced to work at its best. If it is in a state of dysbiosis, this means it is imbalanced or there is a disturbance in the normal composition of microbes; for example, decreased diversity of beneficial microbes.

what does it do?

Dysbiosis may negatively affect different aspects of your health, including your immune system and physical and mental health.

enzyme

what is it?

Digestive enzymes are crucial for good gut health! Your body produces lots of different digestive enzymes to support the digestion and the absorption of fats, proteins, carbohydrates, and micronutrients. For example, amylase is secreted by your salivary glands and pancreas to break carbohydrates down into glucose, which your body can use as fuel, and lipase is secreted by the pancreas to support fat digestion.

what does it do?

Digestive enzymes help to break down our food so we can absorb it. If we don't break down our food properly, it can affect our gut microbes.

feces

what are they?
This is just another word for poo! Feces are made up of water, bacteria, fats, proteins, toxins, and undigested food, including fiber, which helps bulk out your stool.

what do they do?
The purpose of having a poo is to rid your body of waste, toxins, and other compounds.

fiber

what is it?
Fiber is a type of carbohydrate either naturally derived from plants or extracted and added into a product (isolated fiber). Unlike some carbohydrates, fiber cannot be digested in the small intestine and so passes through to the large intestine where the magic happens. It is an essential nutrient for your gut and your gut microbes to thrive.

There are many different types of fiber. You'll often here the term "soluble and insoluble"–this just refers to whether it can be dissolved or not in water. Splitting up fiber in this way doesn't really tell us much about the effect it has on the body and your gut will differ depending on the types of fiber and proportions found in any given food.

what does it do?

Fiber has many important uses in the body.

• Fiber bulks out and softens your stool by retaining water, which supports gut transit time and prevents constipation.
• Certain types of fiber can be fermented by beneficial gut bacteria, which produce short chain fatty acids, which are a source of energy and also have other health benefits.
• Fiber slows the breakdown of sugars found in carbohydrates, which helps to stabilize your energy levels.
• Fiber promotes an environment favorable to beneficial gut bacteria.
• A diet high in fiber can reduce the risk of developing high cholesterol, heart disease, diabetes, and bowel cancer.

Gut associated lymphoid tissue (GALT)

what is it?

Yes, it is a bit of a mouthful, but GALT is extremely important and a huge part of your immune system! It consists of lots of immune cells lining your gut and it plays a fundamental role in defending your body against foreign invaders.

what does it do?

Protects you from pathogens such as bacteria and viruses. Your gut microbiota can influence how well your GALT works and vice versa.

gut

what is it?

When we talk about the gut, we mean your small and large intestine. Your small intestine has an average length of 10–20 feet! Your large intestine (where the majority of your gut bugs live) is around 5 feet long and makes up one fifth of your digestive tract. Your large intestine houses most of your gut bacteria. Your gut is supported by your stomach, liver, pancreas, and gallbladder.

what does it do?

Your gut has many functions, but the key ones are:

• houses the majority of your microbiome
• digests and absorbs nutrients
• supports immune function
• plays a role in producing chemicals that affect how we feel

In your lifetime, around 60 tonnes of food will pass through your gut!

gut microbiome

what is it?

The trillions of microbes, their functions and genes (including bacteria, yeast, fungi, and parasites and their genetic material) living within your gut. More than 1,000 species have been identified! See our infographic on page 27.

what does it do?

Our microbiome does incredible things—here are our top pics:

- Produces vitamins, including vitamin K and B vitamins
- Produces short chain fatty acids, which fuel your gut cells
- Supports immune function and defends against pathogens
- Influences how often you "go" to the loo
- Regulates the health of your gut
- Ferments fiber that your body cannot digest
- Influences mood and mental health
- Influences sleep
- Supports hormone regulation
- Regulates metabolism

gut microbiota

what is it?

The types of organisms (bacteria, viruses, parasites etc.) present in your gut. You might also hear the terms "microbiota" or "microflora" used interchangeably, but we use the term microbiota.

Diet, medication, environment, and genes are just some of the factors that can influence your gut microbiota.

Our microbiomes are unique to each and every one of us! Twins may have the same DNA, but their microbiomes will be different.

Our gut bugs can determine how well we manage different foods, from bananas to avocados.

what does it do?

Your microbiota is responsible for the following:

- Produces vitamins, including vitamin K and B vitamins
- Produces short chain fatty acids, which fuel your gut cells
- Supports immune function and defends against pathogens
- Influences how often you "go" to the loo
- Regulates the health of the gut

- Ferments fiber that your body cannot digest
- Influences mood and mental health
- Influences sleep
- Supports hormone regulation
- Regulates metabolism

Your gut microbes interact with almost all of your human cells!

gut mucosa

what is it?

The gut mucosa includes GALT (see page 153) as well as other immune cells and bacteria, which all work together to distinguish friend from foe in the gut. It allows dietary substances to cross into your bloodstream but still stops pathogens form entering. Specific types of bacteria like to live in your gut mucosa, like our old friend *Akkermansia*.

what does it do?

An important part of our immune system, which provides a balance of beneficial bacteria and stops pathogens from getting in.

polyphenols

what are they?

Polyphenols are protective compounds found in plants; typically they are found in higher quantities in brightly colored vegetables. Your gut microbiota converts polyphenols from plants into something your body can use.

Food sources include brightly colored vegetables, fruit, tea, coffee, chocolate (at least 75% cocoa solids), legumes, and some grains.

what do they do?

Polyphenols have antioxidant properties, support your microbes to be their best selves, and support overall health.

prebiotics

what are they?

Prebiotics are a specific type of fiber, which can be broadly defined as nondigestible carbohydrates. Our gut bacteria like to ferment them, which can cause changes to our gut microbes.

The following are types of prebiotic fiber you might hear about:

- Inulin
- Oligofructose
- Galacto-oligosaccharides (GOS)
- Xylooligosaccharides (XOS)

Great sources of prebiotic foods include onions, garlic, leeks, bananas, asparagus, artichokes, olives, plums, and apples, plus whole grains like oats and bran, and nuts such as almonds.

what do they do?

Prebiotic foods can change the composition of your gut microbes by stimulating growth of beneficial bacteria.

probiotics

what are they?

The term "probiotic" is banded around a lot! The true meaning is a live microorganism that, when administered in adequate amounts, confers a health benefit on the host. Probiotics can be in food or supplement form, but not all probiotics are created equal—different strains have different effects, and some might have no effect at all. It all depends on the individual.

Examples of food probiotics include yogurt, kimchi, sauerkraut, kefir, miso, and kombucha. BUT if you are buying store-bought versions, make sure you check for "live bacteria" on the back of the pack.

what do they do?

Depending on the strain and the individual, probiotics can have lots of positive effects on the person consuming them. Some examples of what they can do:

• compete with other microbes in your gut
• effect your gut mucosal barrier
• support your immune system

Science is still learning exactly how different strains work, so watch this space.

short chain fatty acids

what are they?

Most short chain fatty acids are produced within your large intestine following the fermentation of fiber by your gut microbes.

The main types are:
Acetate
Propionate
Butyrate

Each have different functions in the body.

Fermented products may also contain short chain fatty acids. If we don't eat enough fiber, this can decrease the amount of "food" that bacteria have to ferment and therefore reduce the number of short chain fatty acids produced.

what do they do?

Short chain fatty acids keep your gut healthy: they are the primary source of energy for gut cells, are involved in the metabolism of nutrients (including carbohydrates and fat), support your immune system, and may be protective against certain diseases.

We are still learning about the role of short chain fatty acids and health.

gut conditions

Candida

Candida is a type of fungus or yeast that grows all over the human body, especially in warm and moist areas like the mouth, stomach, and vagina. The presence of candida isn't usually a problem unless an overgrowth occurs. Although overgrowth is quite common, the severity of these infections varies greatly. Candida overgrowth is also called: Candidiasis, a Candida infection, a yeast infection, a fungal infection, and thrush.

Celiac disease

Celiac disease is an autoimmune condition whereby your body's own immune system attacks itself when gluten is eaten (even very small amounts like in the case of cross contamination). This causes damage to your small intestine and can result in malabsorption, meaning you won't be able to absorb the nutrients from food. Your doctor is always your first port of call if you think you have celiac disease.

Clostridium difficile (C. *diff*, as it's known to its pals)

You may have heard of this bacterium. It can infect the bowel and cause diarrhea (in healthy individuals it is unlikely to cause an issue because beneficial bacteria keep it in check). Symptoms include diarrhea, high temperature, or tender stomach.

Crohn's disease

Crohn's disease is a type of inflammatory bowel disease. It can cause inflammation in any part of the gut but mostly occurs in the last section of the small intestine or large intestine. It is a chronic condition but periods of remission mean the individual can be symptom free at certain times. It can impair digestion and absorption of nutrients and how well your body gets rid of waste. Symptoms vary as different sections of the gut are affected. The main symptoms are diarrhea, stomach aches/cramps, blood in your stools, tiredness, and weight loss. Symptoms can be managed with medication and in some cases, surgery.

Diverticular disease and Diverticulitis

These are both conditions affecting your large intestine. As we age, small pockets appear in the lining of our large intestine—these are called diverticula. The presence of diverticula does not mean you will experience symptoms, but some people do. If you have no symptoms but have diverticula, it is called diverticulosis. If you have symptoms, such as stomach pain, it is called diverticular disease. Sometimes the diverticula can become infected or inflamed, increasing the severity of symptoms—this is called diverticulitis.

Digestive cancers

Digestive cancers include:
Colorectal cancer, also known as bowel cancer, which develops
   in the large intestine or rectum.
Gastric cancer, where cells form in the stomach lining.
Pancreatic cancer, where cells form in pancreatic cells.
Other cancers, considered rare, affecting the digestive tract.

Dumping syndrome

Dumping syndrome is a term used to describe a range of different symptoms that occurs when food is too quickly evacuated from the stomach into the small intestine, resulting in food that hasn't been digested properly, making it difficult to absorb nutrients. It has different causes and is common after bariatric surgery (or bypass surgery).

Esophagitis

Esophagitis is where the esophagus (linking your stomach to your throat) becomes inflamed. This mostly occurs due to acid reflux.
Not everyone with reflux will develop esophagitis; it depends on the sensitivity of your esophagus. Symptoms are similar to acid reflux, including pain in the chest and toward the neck, an acid taste in the mouth, and pain when swallowing hot drinks.

Gallstones

Most gallstones consist of cholesterol and are made in your gallbladder. They often go unnoticed but sometimes they can get trapped trying to get out, causing intense stomach pain.

Gastrointestinal malabsorption

Gastrointestinal malabsorption is where you are unable to fully absorb nutrients from your gut, which, if the underlying cause is not treated or managed, can cause malnutrition. The most common cause is celiac disease, Crohn's disease, and pancreatitis. Symptoms include weight loss, impaired growth in children, chronic diarrhea, fatty and greasy stools, and fatigue.

Gastroparesis

Gastroparesis is a chronic condition causing the contents of the stomach to empty slower than it should. It is caused by an impairment in muscular and nerve communication. There are multiple reasons behind its cause, including complications with type 1 and 2 diabetes, Parkinson's disease, and complications from surgery, such as bariatric surgery and gastrectomy. Symptoms include feeling very full (quickly), nausea and vomiting, loss of appetite, weight loss, bloating and stomach discomfort, and heartburn.

H. pylori

Helicobacter pylori is a type of bacterium. An *H. pylori* infection is a bacterial infection of the stomach, where the bacterium sets up home in the mucus layer of your stomach. It weakens the mucus and exposes the stomach lining; the bacteria then irritate the lining, resulting in an ulcer and inflammation. Some people may have an *H. pylori* infection but not experience symptoms. Common symptoms include abdominal pain, bloating, belching, and nausea. It can be treated but requires medical attention.

Heartburn and reflux

Heartburn is a burning sensation in the chest caused by stomach acid or its contents being regurgitated and coming up into the esophagus or throat (acid reflux). It commonly occurs after meals and when lying down. Up to 25% of UK adults are affected. Symptoms vary but include sore throat, heartburn, indigestion, a bad taste, and the sensation of a lump in your throat. If this occurs regularly, its proper name is gastroesophageal reflux disease (GERD). There are different underlying causes, including the sphincter at the bottom of your esophagus/top of your stomach not working as well as it should.

Hemorrhoids (piles)

These are more common than you'd think. Hemorrhoids are swollen blood vessels and appear as lumps in or around your bottom. You are more likely to get them if you are constipated, strain too hard, during pregnancy, and even heavy lifting. They can be painful and may result in bright red blood after you poo, but always seek help from your doctor to rule out other causes.

Hernia

There are several different types of hernia. They occcur when an internal part of your body is pushed through muscle or tissue wall. The most common place is between the chest and hips. If you have a hernia, you will most likely notice a lump or swelling, and surgery may be recommended to treat it. If you think you have a hernia, please speak to your doctor.

IBS

Irritable bowel syndrome affects your digestive system. Symptoms include stomach cramps, diarrhea, constipation, bloating, wind, mucus, nausea, and tiredness—there are different types and your doctor will diagnose you. Symptoms can happen sporadically and last for days, weeks, and even months. Scientists don't know exactly what causes IBS and it can be a difficult condition to live with. Symptoms can be managed but there is no cure. Keep a diary and speak to your doctor for further information.

Intestinal ischemia

This is a rare circulatory condition and can affect your small or large intestine. It occurs where the arteries delivering blood to your gut are affected, resulting in a reduced blood flow.

Pancreatitis

This occurs when your pancreas becomes inflamed, either in a short space of time (acute pancreatitis) or over a long period of time, becoming permanently damaged (chronic pancreatitis). Alcohol abuse is a common cause of chronic pancreatitis.

Short bowel syndrome

Short bowel syndrome happens due to the physical loss or loss of function of a section of the small and/or large intestine. This may result in malabsorption of nutrients. Diarrhea and malnutrition are common symptoms.

SIBO

SIBO stands for small intestinal bacterial overgrowth. This occurs when an unusually high amount of bacteria is found in the small intestine, which leads to uncomfortable symptoms in the gut. This can happen when there is an overgrowth of bacteria in the small intestine, or when bacteria moves from the large intestine into the small intestine.

Stomach/gastric/duodenal ulcers

Stomach or gastric ulcers are ulcerations that occur on the stomach lining or first section of the small intestine. You may also hear the term "peptic ulcer." It happens when the lining of the stomach breaks down, causing damage. It can be caused by an *H. pylori* infection or nonsteroidal anti-inflammatory drugs like aspirin. Burning or abdominal pain are the main symptoms, but you may also experience indigestion, heartburn, or nausea. Seek urgent medical help if you vomit blood, pass dark, sticky stools, have blood in your stools, or sharp pain that gets worse.

Ulcerative colitis

Ulcerative colitis (UC) is a type of inflammatory bowel disease (you may hear it called IBD). It causes inflammation and ulceration of the large intestine and rectum (where stools are held before being released). The large intestine has a delicate lining and ulcers can form along it, which can cause bleeding and pus. Severity of symptoms will vary person to person but the main symptoms are diarrhea (you may see blood, mucus or pus), stomach pain, frequent trips to the bathroom, fatigue, loss of appetite and weight loss, and anemia. Other symptoms are also associated with UC. It is an ongoing condition but symptoms can be managed with medication and, in some cases, surgery.

Seek help from your doctor if you are experiencing gut symptoms.

references

MAIN BODY

Petra Louis, Harry J. Flint, Diversity, metabolism and microbial ecology of butyrate-producing bacteria from the human large intestine, *FEMS Microbiology Letters*, Volume 294, Issue 1, May 2009, Pages 1–8, https://doi.org/10.1111/j.1574-6968.2009.01514.x

Zinöcker, M. K., & Lindseth, I. A. (2018). The Western Diet-Microbiome-Host Interaction and Its Role in Metabolic Disease. *Nutrients, 10*(3), 365. https://doi.org/10.3390/nu10030365

https://assets.publishing.service. gov.uk/government/uploads/ system/uploads/attachment_data/ file/445503/SACN_Carbohydrates_ and_Health.pdf

Myhrstad, M., Tunsjø, H., Charnock, C., & Telle-Hansen, V. H. (2020). Dietary Fiber, Gut Microbiota, and Metabolic Regulation-Current Status in Human Randomized Trials. *Nutrients, 12*(3), 859. https://doi.org/10.3390/ nu12030859

Sanders, M.E., Merenstein, D.J., Reid, G. et al. Probiotics and prebiotics in intestinal health and disease: from biology to the clinic. *Nat Rev Gastroenterol Hepatol* 16, 605–616 (2019). https://doi.org/10.1038/ s41575-019-0173-3

Kumar Singh, A., Cabral, C., Kumar, R., Ganguly, R., Kumar Rana, H.,

Gupta, A., Rosaria Lauro, M., Carbone, C., Reis, F., & Pandey, A. K. (2019). Beneficial Effects of Dietary Polyphenols on Gut Microbiota and Strategies to Improve Delivery Efficiency. *Nutrients, 11*(9), 2216. https://doi.org/10.3390/nu11092216

American Gut: an Open Platform for Citizen Science Microbiome Research

Daniel McDonald, Embriette Hyde, Justine W. Debelius, James T. Morton, Antonio Gonzalez, Gail Ackermann, Alexander A. Aksenov, Bahar Behsaz, Caitriona Brennan, Yingfeng Chen, Lindsay DeRight Goldasich, Pieter C. Dorrestein, Robert R. Dunn, Ashkaan K. Fahimipour, James Gaffney, Jack A. Gilbert, Grant Gogul, Jessica L. Green, Philip Hugenholtz, Greg Humphrey, Curtis Huttenhower, Matthew A. Jackson, Stefan Janssen, Dilip V. Jeste, Lingjing Jiang, Scott T. Kelley, Dan Knights, Tomasz Kosciolek, Joshua Ladau, Jeff Leach, Clarisse Marotz, Dmitry Meleshko, Alexey V. Melnik, Jessica L. Metcalf, Hosein Mohimani, Emmanuel Montassier, Jose Navas-Molina, Tanya T. Nguyen, Shyamal Peddada, Pavel Pevzner, Katherine S. Pollard, Gholamali Rahnavard, Adam Robbins-Pianka, Naseer Sangwan, Joshua Shorenstein, Larry Smarr, Se Jin Song, Timothy Spector, Austin D. Swafford, Varykina G. Thackray, Luke R. Thompson, Anupriya Tripathi, Yoshiki Vázquez-Baeza, Alison Vrbanac, Paul Wischmeyer, Elaine Wolfe, Qiyun Zhu, The American Gut Consortium, Rob Knight mSystems May 2018, 3 (3) e00031-18; DOI: 10.1128/mSystems.00031-18

Karl, J. P., Hatch, A. M., Arcidiacono, S. M., Pearce, S. C., Pantoja-Feliciano, I. G., Doherty, L. A., & Soares, J. W. (2018). Effects of Psychological, Environmental and Physical Stressors on the Gut Microbiota. *Frontiers in microbiology, 9*, 2013. https://doi.org/10.3389/fmicb.2018.02013

Zheng, D., Liwinski, T. & Elinav, E. Interaction between microbiota and immunity in health and disease. *Cell Res 30*, 492–506 (2020). https://doi.org/10.1038/s41422-020-0332-7

Martin, A. M., Sun, E. W., Rogers, G. B., & Keating, D. J. (2019). The Influence of the Gut Microbiome on Host Metabolism Through the Regulation of Gut Hormone Release. *Frontiers in physiology, 10*, 428. https://doi.org/10.3389/fphys.2019.00428

Rizzello, C. G., Portincasa, P., Montemurro, M., Di Palo, D. M., Lorusso, M. P., De Angelis, M., Bonfrate, L., Genot, B., & Gobbetti, M. (2019). Sourdough Fermented Breads are More Digestible than Those Started with Baker's Yeast Alone: An In Vivo Challenge Dissecting Distinct Gastrointestinal Responses. *Nutrients, 11*(12), 2954. https://doi.org/10.3390/nu11122954

Koistinen, V. M., Mattila, O., Katina, K., Poutanen, K., Aura, A. M., & Hanhineva, K. (2018). Metabolic profiling of sourdough fermented wheat and rye bread. *Scientific reports, 8*(1), 5684. https://doi.org/10.1038/s41598-018-24149-w

Reese, A. T., Madden, A. A., Joossens, M., Lacaze, G., & Dunn, R. R. (2020). Influences of Ingredients and Bakers on the Bacteria and Fungi in Sourdough Starters and Bread. *mSphere, 5*(1), e00950-19. https://doi.org/10.1128/mSphere.00950-19

Melini, F., Melini, V., Luziatelli, F., Ficca, A. G., & Ruzzi, M. (2019). Health-Promoting Components in Fermented Foods: An Up-to-Date Systematic Review. *Nutrients, 11*(5), 1189. https://doi.org/10.3390/nu11051189

Bourrie, B. C., Willing, B. P., & Cotter, P. D. (2016). The Microbiota and Health Promoting Characteristics of the Fermented Beverage Kefir. *Frontiers in microbiology, 7*, 647. https://doi.org/10.3389/fmicb.2016.00647

Lobionda, S., Sittipo, P., Kwon, H. Y., & Lee, Y. K. (2019). The Role of Gut Microbiota in Intestinal Inflammation with Respect to Diet and Extrinsic Stressors. *Microorganisms, 7*(8), 271. https://doi.org/10.3390/microorganisms7080271

Davani-Davari, D., Negahdaripour, M., Karimzadeh, I., Seifan, M., Mohkam, M., Masoumi, S. J., Berenjian, A., & Ghasemi, Y. (2019). Prebiotics: Definition, Types, Sources, Mechanisms, and Clinical Applications. *Foods (Basel, Switzerland), 8*(3), 92. https://doi.org/10.3390/foods8030092

Carlson, J. L., Erickson, J. M., Lloyd, B. B., & Slavin, J. L. (2018). Health Effects and Sources of Prebiotic Dietary Fiber. *Current developments in nutrition, 2*(3), nzy005. https://doi.org/10.1093/cdn/nzy005

Househam, A. M., Peterson, C. T., Mills, P. J., & Chopra, D. (2017). The Effects of Stress and Meditation on the Immune System, Human Microbiota, and Epigenetics. *Advances in mind-body medicine, 31*(4), 10–25.

MENTAL HEALTH
Long-Smith, C., O'Riordan, K. J., Clarke, G., Stanton, C., Dinan, T. G., & Cryan, J. F. (2020). Microbiota-Gut-Brain Axis: New Therapeutic Opportunities. *Annual review of pharmacology and toxicology, 60*, 477–502. https://doi.org/10.1146/annurev-pharmtox-010919-023628

Yang B, Wei J, Ju P, et al., Effects of regulating intestinal microbiota on anxiety symptoms: A systematic review

General Psychiatry
2019;32:e100056. doi: 10.1136/gpsych-2019-100056

[John/Kimberley]

KRISTY COLEMAN
Clark A, Mach N. Exercise-induced stress behavior, gut-microbiota-brain axis and diet: a systematic review for athletes. *J Int Soc Sports Nutr.* 2016;13:43. Published 2016 Nov 24. doi:10.1186/s12970-016-0155-6

Dalton A, Mermier C, Zuhl M. Exercise influence on the microbiome-gut-brain axis. *Gut Microbes.* 2019;10(5):555-568. doi:10.1080/19490976.2018.1562268

Das M, Cronin O, Keohane DM, et al. Gut microbiota alterations associated with reduced bone mineral density in older adults. Rheumatology (Oxford, England). 2019 Dec;58(12):2295-2304. DOI: 10.1093/rheumatology/kez302.

Mailing LJ, Allen JM, Buford TW, Fields CJ, Woods JA. Exercise and the Gut Microbiome: A Review of the Evidence, Potential Mechanisms, and Implications for Human Health. *Exerc Sport Sci Rev.* 2019;47(2):75-85. doi:10.1249/JES.0000000000000183

JENNA MACCHIOCI
Tamburini, S., Shen, N., Wu, H. et al. The microbiome in early life: implications for health outcomes. Nat Med 22, 713–722 (2016). https://doi.org/10.1038/nm.4142

Nuria Salazar, Silvia Arboleya, Tania Fernández-Navarro, Clara G. de los Reyes-Gavilán,Sonia Gonzalez, and Miguel Gueimonde. Age-Associated Changes in Gut Microbiota and Dietary Components Related with the Immune System in Adulthood and Old Age: A Cross-Sectional Study Nutrients. 2019 Aug; 11(8): 1765. doi: 10.3390/nu11081765

Mowat AM. To respond or not to respond - a personal perspective of intestinal tolerance [published correction appears in Nat Rev Immunol. 2018 Aug;18(8):536]. Nat Rev Immunol. 2018;18(6):405-415. doi:10.1038/s41577-018-0002-x

McDonald, D., Hyde, E., Debelius, J. W., Morton, J. T., Gonzalez, A., Ackermann, G., ... Gunderson, B. (2018). American Gut: an Open Platform for Citizen Science Microbiome Research. MSystems,

3(3). https://doi.org/10.1128/msystems.00031-18

Laitinen K., Mokkala K. Overall Dietary Quality Relates to Gut Microbiota Diversity and Abundance. Int. J. Mol. Sci. 2019;20:1835. doi: 10.3390/ijms20081835. [PMC free article] [PubMed] [CrossRef] [Google Scholar]

Rinninella E, Raoul P, Cintoni M, Franceschi F, Miggiano GAD, Gasbarrini A, et al. What is the healthy gut microbiota composition? A changing ecosystem across age, environment, diet, and diseases. Microorganisms. 2019;7(1). 10.3390/microorganisms7010014. [PMC free article] [PubMed]

Maslowski KM, Mackay CR. Diet, gut microbiota and immune responses. Nat Immunol. 2011;12(1):5-9. doi:10.1038/ni0111-5

Dimidi E, Cox SR, Rossi M, Whelan K. Fermented Foods: Definitions and Characteristics, Impact on the Gut Microbiota and Effects on Gastrointestinal Health and Disease. Nutrients. 2019;11(8):1806. Published 2019 Aug 5. doi:10.3390/nu11081806

Jacob G. Mills, Justin D. Brookes, Nicholas J. C. Gellie, Craig Liddicoat, Andrew J. Lowe, Harrison R. Sydnor, Torsten Thomas, Philip Weinstein, Laura S. Weyrich and Martin F. Breed. Relating Urban Biodiversity to Human Health With the 'Holobiont' Concept. Front. Microbiol., 26 March 2019 | https://doi.org/10.3389/fmicb.2019.00550

Cox L. M., Blaser M. J. (2015). Antibiotics in early life and obesity. Nat. Rev. Endocrinol. 11:182. 10.1038/nrendo.2014.210 [PMC free article] [PubMed]

SOPHIE MEDLIN
Patel SG, Ahnen DJ. Colorectal Cancer in the Young. *Curr Gastroenterol Rep.* 2018;20(4):15. Published 2018 Mar 28. doi:10.1007/s11894-018-0618-9

Flynn S, Eisenstein S. Inflammatory Bowel Disease Presentation and Diagnosis. *Surg Clin North Am.* 2019;99(6):1051-1062. doi:10.1016/j.suc.2019.08.001

Lacy BE, Patel NK. Rome Criteria and a Diagnostic Approach to Irritable Bowel Syndrome. *J Clin Med.* 2017;6(11):99. Published 2017 Oct 26. doi:10.3390/jcm6110099

Lebwohl B, Sanders DS, Green PHR. Coeliac disease. *Lancet.* 2018;391(10115):70-81. doi:10.1016/S0140-6736(17)31796-8

Vijayvargiya P, Camilleri M. Update on Bile Acid Malabsorption: Finally Ready for Prime Time?. *Curr Gastroenterol Rep.* 2018;20(3):10. Published 2018 Mar 26. doi:10.1007/s11894-018-0615-z

van der Heide F. Acquired causes of intestinal malabsorption. *Best Pract Res Clin* Gastroenterol. 2016;30(2):213-224. doi:10.1016/j.bpg.2016.03.001

Rezapour M, Ali S, Stollman N. Diverticular Disease: An Update on Pathogenesis and Management. *Gut Liver.* 2018;12(2):125-132. doi:10.5009/gnl16552

CHRIS GEORGE

Pechey, R., & Monsivais, P. (2016). Socioeconomic inequalities in the healthiness of food choices: Exploring the contributions of food expenditures. *Preventive medicine, 88*, 203–209. https://doi.org/10.1016/j.ypmed.2016.04.012

https://www.who.int/nutrition/topics/2_background/en/

https://www.who.int/nutrition/topics/2_background/en/

https://www.who.int/gho/ncd/mortality_morbidity/en/

https://www.who.int/chp/about/integrated_cd/en/

https://ueg.eu/a/43

LAURA TILT

Simren, M., Lembo, A. J., Chang, L., Mearin, F., Chey, W. D., Lacy, B. E., & Spiller, R. (2016). Bowel Disorders. Gastroenterology, *150*(6), 1393–1407. https://doi.org/10.1053/j.gastro.2016.02.031

McKenzie, Y. A., Bowyer, R. K., Leach, H., Gulia, P., Horobin, J., O'Sullivan, N. A., Pettitt, C., Reeves, L. B., Seamark, L., Williams, M., Thompson, J., & Lomer, M. C. E. (2016). British Dietetic Association systematic review and evidence-based practice guidelines for the dietary management of irritable bowel syndrome in adults (2016 update). *Journal of Human Nutrition and Dietetics : The Official Journal of the British Dietetic Association.* https://doi.org/10.1111/jhn.12385

National Institute for Health and Care Excellence (NICE). (2008). *Irritable bowel syndrome in adults: diagnosis and management. Clinical guideline.* https://www.nice.org.uk/guidance/cg61/

Drossman, D. A., Morris, C. B., Schneck, S., Hu, Y. J. B., Norton, N. J., Norton, W. F., Weinland, S. R., Dalton, C., Leserman, J., & Bangdiwala, S. I. (2009). International survey of patients with IBS: Symptom features and their severity, health status, treatments, and risk taking to achieve clinical benefit. *Journal of Clinical Gastroenterology, 43*(6), 541–550. https://doi.org/10.1097/MCG.0b013e318189a7f9

Drossman, D. A. (2016). Functional gastrointestinal disorders: History, pathophysiology, clinical features, and Rome IV. *Gastroenterology, 150*(6), 1262-1279.e2. https://doi.org/10.1053/j.gastro.2016.02.032

bibliography

Tim Spector
Spoon-fed (Jonathan Cape, 2020)
The Diet Myth (Weidenfeld & Nicolson, 2016)

Jenna Macciochi
Immunity (Harper Collins, 2020)

Kimberley Wilson
How to Build a Healthy Brain (Yellow Kite, 2020)

Anjali Mahto
The Skincare Bible (Penguin Life, 2018)

Renee Mcgregor
Orthorexia (Nourish Books, 2017)
Fast Fuel (Nourish Books, 2016)
Training Food (Nourish Books, 2015)
www.reneemcgregor.com

Scott C. Anderson with John Cryan and Ted Dinan
The Psychobiotic Revolution (National, Geographic, 2017)

Rosie Saunt and Helen West
Is Butter a Carb? (Piatkus, 2019)

Miguel Toribio Mateas
Harnessing the Power of Microbiome Assessment Tools as Part of Neuroprotective Nutrition and Lifestyle Medicine Interventions.
https://www.mdpi.com/2076-2607/6/2/35

WEBSITES & RESOURCES
Dr Rabia
@doctor_rabia
www.doctor-rabia.com

Sophie Medlin
@sophiedietitian

Laura Tilt
@nutritilty

Ruari Robertson
www.ruairirobertson.com

index

A
alcohol 111–13
antibiotics 11, 45
antioxidants 99
anxiety 57
appendix 19

B
babies 44
bacteria 148
  fermented food 94–5
  prebiotics 65–6, 69–73,
    143, 159
  probiotics 45, 51, 65–8,
    143, 160
  SIBO 166
  yogurt 98
  see also gut microbiome
bile 18, 149
birth 44
bloating 127, 130
blood, in poo 130, 133
brain 52–7
bread 94
breathing 118–20
butyrate 46–7, 54, 161

C
caffeine 61
cancer 163
candida 162
canned foods 105
carbohydrates 78, 88
chewing food 84
chocolate 109
chronic diseases 34–6
chyme 16, 18, 149
circadian rhythm 58–9, 61
clean eating 78
Clostridium difficile 140, 162
Coeliac disease 162, 163
condiments 110
constipation 133
Crohn's disease 162, 163

D
depression 54, 55
detoxes 76, 81
digestive problems 126–37,
  162–7
diverticula 133
diverticular disease 163
diverticulitis 163
DNA sequencing 142
dopamine 56, 57, 58
drinks 110, 111–13
dumping syndrome 163
duodenal ulcers 166
dysbiosis 28–9, 49, 150

E
eating disorders 114–16
enzymes 16, 17, 18, 113, 150
epithelium 23–4
esophagitis 165
esophagus 16
estrogen 62
exercise 46–7

F
fad diets 78, 81
fasting 86
fecal transplants 140
feces (poo) 19, 121–3,
126–33, 151
fermented food 94–103
fiber 11, 44, 46, 55, 69,
88–93, 137, 151–2, 159
FODMAP foods 116
food: chewing 84
  fridges and freezers 106–7
  larder (pantry) 104–5
  swaps 108–10
food diaries 117
free radicals 99
freezers 106–7, 109
fridges 106–7
fruit 78

G
GABA 56, 57, 58
gallstones 163
gastric ulcers 166
gastrointestinal
malabsorption 163
gastroparesis 164
genetics 142
gut 15–19, 154
gut associated lymphoid
tissue (GALT) 153, 157
gut-brain axis 52–7, 58
gut microbiome 18, 20–2,
26–8, 155
  dysbiosis 28–9, 49, 150
  influences 30–3
  sources of microbes 44, 45
gut microbiota 10–11, 26–8,
156–7
gut mucosa 23–4, 157

H
H. pylori 164
haemorrhoids (piles) 133, 164

hangovers 113
heartburn 164–5
herbs 105
hernia 165
hierarchy of evidence 37
hormones 18, 62–3

I
IBS (irritable bowel syndrome) 134–7, 165
immune system 40–5, 55, 153
inflammation 22, 23, 49, 55, 113
intestinal ischaemia 165
inulin 70–1

K
kefir 95, 102
kimchi 95, 102
kombucha 98

L
large intestine 18, 23–4, 154
leaky gut 55, 113
lifestyles 32, 34–5
liver 17–18

M
magic pills 78
microbes see gut microbiome
microbiota 10–11, 26–8, 156–7
mindful eating 85
miso 95
mouth 16
mucus layer 23–4, 47, 55, 157
myths 79–81

N
nervous system 54, 118, 119
neurotransmitters 57, 58

O
oils 105
orthorexia 114–16

P
pain 127, 130
pancreas 18
pancreatitis 163, 166
parasympathetic nervous system 54, 118
pasta 109
periods 63
personalised treatments 143
piles 133, 164
polyphenols 99, 112, 158
poo 19, 121–3, 126–33, 151
prebiotics 65–6, 69–73, 143, 159
probiotics 45, 51, 65–8, 143, 160

R
RCTs 36–7
reflux 164–5
research 36–7, 102–3
resistant starch 89

S
sauerkraut 94, 96–7, 102
serotonin 54, 56, 57, 58
short bowel syndrome 166
short chain fatty acids 46, 54, 88–9, 161
SIBO 166
sitting position, pooing 121–2

skin 48–51
skinny teas 76
sleep 58–61
small intestine 17, 23–4, 154
soups 110
sourdough 94
spices 105
squatting position, pooing 121, 123
stomach 16–17, 164
stomach ulcers 166
stools (poo) 19, 121–3, 126–33, 151
stress 118–20, 136
superfoods 76
swaps, food 108–10

T
toilets, pooing position 121–3
tongue 16

U
ulcerative colitis (UC) 166–7
ulcers 166

V
vagus nerve 54, 119–20
vegetables 109
vinegars 105

W
water, drinking 92
wind 127, 131
wine 112–13

Y
yogurt 98, 109

acknowledgements

Firstly, to our team at TGS HQ who have stuck by us on this business rollercoaster and everyday give us the incredible combination of heart and graft, we don't know where we'd be without you all.

To India who kicked this baby off with us, and everyone that believed in us from the very beginning, JKR and Revolt who sat in the tiny little sailboat with us as we set off on this journey... and the Haltons for giving us the strength when we thought we might sink.

To Shelly Nel who's warmth behind the camera was the catalyst for us to have access to the best scientists and professionals around.

To Luke and Peter, for the late nights and early mornings—thanks for standing by us.

To Mum who continues to hold our trembling hands when we step out of our comfort zone and to Uncle Brian and all the pals who read very early drafts and scribbles of these pages.

To our incredible community of followers—you are who we do this for.

And to everyone we've yet to empower—here's to the future.

To everyone who said to us all those years ago, what are you two on about?! Do you mean my beer belly? I hope now after reading this book you understand and share our passion to tell others. Onwards!

Lisa and Alana x